ZINC OXIDE NANOSTRUCTURES

ZINC OXIDE NANOSTRUCTURES

ADVANCES AND APPLICATIONS

edited by

Magnus Willander

PAN STANFORD PUBLISHING

Published by

Pan Stanford Publishing Pte. Ltd.
Penthouse Level, Suntec Tower 3
8 Temasek Boulevard
Singapore 038988

Email: editorial@panstanford.com
Web: www.panstanford.com

British Library Cataloguing-in-Publication Data
A catalogue record for this book is available from the British Library.

Zinc Oxide Nanostructures: Advances and Applications

ISBN 978-981-4411-33-2 (Hardcover)
ISBN 978-981-4411-34-9 (eBook)

Printed in the USA

Contents

Preface ix

1. Fabrication of ZnO Nanostructures **1**

Andrey Bakin, Arne Behrends, Alexander Wagner, and Andreas Waag

1.1 Introduction 2
1.2 ZnO Nanostructure Fabrication 3
1.3 Methods of ZnO Nanostructure Fabrication 3
1.4 Wet Chemical Fabrication 4
1.4.1 Chemical Bath Deposition 4
1.4.2 Electrodeposition 8
1.4.3 Successive Ionic Layer Adsorption and Reaction 11
1.5 Gas-Phase Growth 13
1.5.1 Vapor Phase Transport and Chemical Vapor Transport 13
1.5.2 Chemical Vapor Deposition 16
1.6 Pulsed Laser Deposition 21
1.7 ZnO Nanopowder-Based Technologies 22
1.8 Fabrication of Complicated ZnO-Based Heteronanostructures 25
1.8.1 Fabrication of Quantum Wells Embedded into ZnO Nanostructures 25
1.8.2 Combination of ZnO Nanostructures with Other Semiconductors 26
1.9 Summary 33

2. Optical Properties of and Optical Devices from ZnO-Based Nanostructures **43**

Michael Lorenz, Martin Lange, Christian Kranert, Christof P. Dietrich, and Marius Grundmann

2.1 Introduction 43
2.2 Optical Recombination 44
2.2.1 Optical Recombination of ZnO Nanostructures 45
2.2.2 Bandgap Engineering 51
2.2.3 Nanowire-Heterostructures: Quantum Wells and Quantum Dots 55
2.3 Cavity Effects 60
2.3.1 Whispering Gallery Modes 60
2.3.2 Fabry–Perot Modes 65
2.3.3 Random Lasing 68
2.4 Raman and Infrared Spectra from Nanostructures 70
2.5 Nanostructure Devices 75
2.5.1 Light-Emitting Diodes and Lasers 75
2.5.2 Photodetectors 77
2.5.3 Second Harmonic Generation 80
2.6 Summary and Outlook 82

3. Piezoelectric Nanogenerator Based on ZnO Nanomaterials **101**

Jinhui Song

3.1 Introduction: An Explanation of the Motivation for Developing Nanogenerator 102
3.2 First Nanogenerator 102
3.3 Visualizing Power Generation Process 107
3.4 Piezoelectric Potential in NW 108
3.5 Direct Current Nanogenerator 110
3.6 High-Power-Output Nanogenerator 115
3.7 Laterally Packed AC Nanogenerator 118
3.8 High-Output Lateral Nanogenerator 120
3.9 Summary 124

4. Nanobiology and Nanomedical Devices Using Zinc Oxide Nanostructures **127**

Magnus Willander and Omer Nur

4.1 Introduction 128
4.2 Growth and Fabrication of ZnO-Based Electrodes 130
4.3 Extra- and Intracellular Sensing of Chemical Ions 132
4.3.1 Intracellular Ca^{2+} Ion Detection 132
4.3.2 Intracellular Na^{1+} Ion Detection 136
4.4 Extra- and Intracellular Sensing of Glucose 138
4.5 Extra- and Intracellular Sensing of Biological Analytes and Other Chemical Substances 142
4.5.1 Cholesterol Sensing Using ZnO Nanostructures 142
4.5.2 Uric Acid Detection Using ZnO Nanowires 144
4.6 Zinc Oxide Nanowires as Research Tool for Cancer Cell Treatment 147
4.7 Interfacing ZnO Nanowire–Based Sensors with Conventional Electronic Components 150
4.8 Summary 151

5. ZnO Nanostructures: Toxicity and Phototoxicity Characteristics in Biological Samples **157**

Magnus Willander, O. Nur, M. Fakhr-e-Alam, M. Atif, and M. S. AlSalhi

5.1 Cell Culture 162
5.2 Freestanding Drug Delivery 164
5.2.1 Preparation of Metal Oxides Nanostructures 164
5.3 Stock Solution of Chemicals and Photosensitizers 164
5.4 Light Exposure 164
5.5 Drug Cytotoxicity and Phototoxicity Tests 165
5.6 Incubation of Photosensitizer and Laser Irradiation 165
5.7 Labeling of Carcinogenic Cells with MnO_2 NRs, Fe_2O_3 NPs, ZnO NRs and Their Conjugated form with 5-ALA and PPDME 166

5.8 Viability Determination 166
5.9 Reactive Oxygen Species Measurement 167
5.10 Confocal Microscopy 167
5.11 Microinjection Drug Delivery 175
5.12 Deposition of Nanostructures on Borosilicate Glass Capillary (Femtotip®) 175

6. Zinc Oxide Nanostructures: Synthesis, Characterization, and Device Applications on Nonconventional Substrates 185

Magnus Willander, Omer Nur, Kamran ul Hasan, Gul Amin, and Muhammad Yousuf Soomro

6.1 Introduction 186
6.2 Experimental 188
6.3 A Novel Low-Temperature Direct Growth of ZnO Nanorods on Paper Substrates 190
6.4 Results 192
6.5 Chemically Grown ZnO Nanorods Printed on Flexible Plastic and Paper Substrates 197
6.6 Conclusion 206

Index 213

Preface

Zinc oxide (ZnO) is a semiconducting material with more interesting and important properties than most other materials. This has resulted in a vast number of scientific publications on ZnO, particularly since 2000. The reason for this is that ZnO can be realized very easily in many different nanostructures. In this book, we will restrict ourselves to the realization of nanowires, nanotubes, dots, and quantum wells.

The first chapter, by A. Bakin et al. from Technische Universität Braunschweig, Germany, describes different methods to fabricate ZnO nanostructures.

The optical properties of ZnO have been of special interest for many decades. In Chapter 2, M. Lorenz's group from Universität Leipzig, Germany, gives a detailed description of the optical properties of dots, wires (and tubes), and quantum wells.

Since 2006, the piezoelectric properties of the ZnO nanowires have been of special interest owing to the first demonstration of the ZnO nanogenerator, which converts mechanical energy to electrical energy. In Chapter 3, J. H. Song describes in detail the physics and technology of this nanogenerator, which eventually will provide a renewable and green energy source.

Nanostructures are becoming very important for chemical and biological sensing. Here zinc oxide nanostructures play an important role. In Chapter 4, M. Willander and O. Nur from Linköping University, Sweden, illustrate this in by their work on metal ion sensing, glucose sensing with potentiometric measurements, intra- and extracellular sensing, and interfacing the signal for transmission via mobile telephone. Also, it is described how owing to their optical spectra, ZnO nanowires can be used to inject singlet oxygen into different parts of a cell and how the cell reacts.

Toxicity of nanomaterials has been investigated and discussed a lot for several years. In Chapter 5, the group from Linköping and Riyadh discuss the phototoxicity of zinc oxide.

Finally, Chapter 6, by M. Willander et al., discusses one of the most interesting properties of zinc oxide nanostructures, i.e., the possibility to grow crystalline ZnO nanostructures and devices (LEDs, detectors, and nanogenerators) or print them on paper. It also discusses the growth on plastic substrates.

Magnus Willander

Chapter 1

Fabrication of ZnO Nanostructures

Andrey Bakin, Arne Behrends, Alexander Wagner, and Andreas Waag

Institute of Semiconductor Technology, Technische Universität Braunschweig, Hans-Sommer-Strasse 66, Braunschweig, Germany

A.Bakin@tu-bs.de

Owing to its excellent luminescent properties, zinc oxide is a promising material for the development of optoelectronic and photovoltaic devices. ZnO epitaxial layer quality and control of native and dopant point defects is still an issue for direct device production. One of the challenges for high-quality epitaxial layer growth is the lack of high-quality, large and affordable ZnO wafers. The implementation of nanostructures is a promising solution. For instance, high-quality nanopillars with a small footprint on the wafer can be deposited on different types of wafers, including flexible ones. This chapter gives a short overview of the methods of the fabrication of ZnO nanostructures. The focus is not on the properties of the ZnO nanostructures but on the growth approaches and their comparison promoting the choice of the proper growth method for certain applications. The methods presented can be classified in three groups: wet chemical approaches, gas-phase approaches, and

Zinc Oxide Nanostructures: Advances and Applications
Edited by Magnus Willander

ISBN 978-981-4411-33-2 (Hardcover), 978-981-4411-34-9 (eBook)
www.panstanford.com

implementation of ZnO nanopowders. The recent advances in this area are presented. The most important aspects of manufacturing both quantum wells (QWs) and p–n junctions embedded into nanostructures are also discussed.

1.1 Introduction

Zinc oxide is a II–VI semiconductor with a wide direct bandgap of 3.3 eV at room temperature. It has a large exciton binding energy [1], which provides ultraviolet (390 nm) excitonic laser action upon optical pumping at room temperature [2]. Therefore, it is considered suitable for optoelectronic applications. The large breakdown strength and saturation velocity also make ZnO a potential material for high-temperature, high-power electronics [3]. High-quality epitaxial thin films are necessary in order to utilize the aforementioned properties of ZnO. Significant progress in ZnO homo- and heteroepitaxy technology has been observed in recent years, but further improvement of ZnO epitaxial layer quality is still necessary for achieving reproducible p-type doping and to realize reliable ZnO-based devices [4]. An alternative solution could be the implementation of nanostructures. Owing to a small footprint on the wafer, high-quality nanopillars can be deposited on different types of wafers, including flexible ones.

In this chapter, we present the state of the art of ZnO nanostructure fabrication technologies, comparing the ZnO nanostructure fabrication techniques, highlighting the most important ones, and giving a very short overview of our results as well as of the results published by various groups of the ZnO community. The objective is to provide information on ZnO nanostructure fabrication and make it easier to address the issues and challenges according to the requirements of certain application. The combination of ZnO with other semiconductor compounds is of special interest since at present the production of reliable p-type ZnO and ZnO-based diluted magnetic semiconductors is still a challenge. From this point of view, it is of special interest to combine ZnO with GaN or other p-type semiconductors for LED or solar cell fabrication. The fabrication of ZnO heteronanostructures is also described in this chapter.

1.2 ZnO Nanostructure Fabrication

Most semiconductor-based optoelectronic and many microelectronic devices are based on heterostructures that consist of layers of different materials with the same crystal structure. In this case, advanced epitaxial growth techniques are needed. However, even in the case of well-developed epitaxial growth technologies, device parameters can be degraded if the substrate material quality or processing is poor. Therefore, substrate fabrication technology is equally important for state-of-the-art semiconductor device manufacturing. Currently, the quality of ZnO bulk material and its surface preparation are not yet mature for direct device production; therefore, ZnO epitaxial growth plays a key role in ZnO device technology. Recently, ZnO nanostructures, such as nanopillars, have become a field of intensive research for ZnO device applications. The main growth approaches are gas-phase growth and chemical solution growth. A comparison of chemical solution and gas-phase growth techniques shows that chemical solution methods provide low-cost fabrication of nanostructures on large wafers and could be easily the method of choice for large-scale production. The gas-phase growth techniques commonly provide the best quality layers and nanostructures, as well as device heterostructures that cannot be realized by chemical solution methods. ZnO films and nanostructures can be manufactured from the gas phase by chemical vapor deposition (CVD), sublimation growth or physical vapor transport (PVT), molecular beam epitaxy (MBE), liquid phase epitaxy (LPE), sputtering, pulsed laser deposition (PLD), atomic layer deposition (ALD), chemical bath deposition (CBD), successive ion layer adsorption reaction (SILAR), electrodeposition (ED), powder deposition, pressing and sintering, spraying, and pyrolysis.

1.3 Methods of ZnO Nanostructure Fabrication

The main ZnO nanostructure fabrication methods described in the chapter are presented in Fig. 1.1. Chemical vapor deposition, chemical vapor transport (CVT), PVT, and PLD are the main gas-phase methods of ZnO nanostructure fabrication. Molecular beam epitaxy and radio frequency (RF) magnetron sputtering are typically used only for the fabrication of QWs embedded into the nanostructures as an additional second step of the fabrication technology employing nanostructures as a substrate, although there are also reports on direct fabrication of ZnO nanorods employing RF magnetron sputtering [69]. The main chemical solution techniques of ZnO nanostructure fabrication are CBD and ED. We also describe here the SILAR method, which was developed for the fabrication of layers [5], first for photovoltaic applications [6] and also nanostructured ZnO for fabrication of sensors [7]. We have demonstrated SILAR growth of core–shell nanostructures employing this method, which is normally difficult to realize from chemical solution.

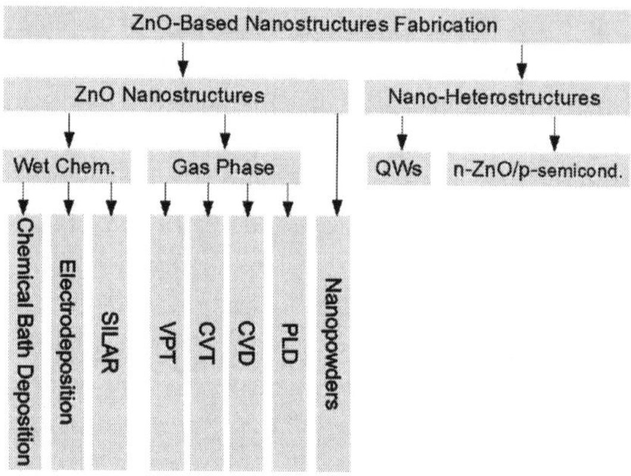

Figure 1.1 The main ZnO nanostructures fabrication methods.

A separate nanotechnological approach is processing of ZnO nanopowders. It is especially promising for solar cells and sensors fabrication.

1.4 Wet Chemical Fabrication

1.4.1 Chemical Bath Deposition

A cost-effective alternative to gas-phase growth is wet chemical deposition, and the most well-known and a widely used wet chemical deposition method is the so-called CBD. In this case, it is not necessary to provide vacuum and high process temperatures. This leads to significantly simpler and cheaper growth setups. Such wet chemical deposition can be typically realized in one or more steps. The whole deposition process occurs in one bath. The anion and the cation of the required compound meet each other in the solution. Typically, metal ions exist in the solution already at room temperatures, and it is necessary to increase the temperature to produce anions in the solution. As a result, the growth of the required end product occurs on the surface of the substrate, which serves partly as the catalysator of this growth.

Temperature-controlled chemical bath reactor schematically presented in Fig. 1.2 was employed for ZnO nanostructure fabrication under ambient pressure. Substrate is placed into the thermally stabilized reactor with an aqueous solution consisting of zinc nitrate ($Zn(NO_3)_2$) and hexamethyleneteramine (HMT, $C_6H_{12}N_4$) at a temperature from 60 to 90°C [8, 9]. Temperature and pH value of the solution are controlled during the growth. Homogeneous distribution of the solution concentration in the reactor is provided by stirring the solution. Duration of ZnO growth is typically several hours. Implementation of a blank substrate leads to the formation of low-density ZnO structures like "flowers" (cf. Fig. 1.3a). Pretreatment of the substrate; formation of a nucleation layer is necessary to promote the formation of high-density ZnO nanorods; so CBD of ZnO is typically realized in two steps [10–14]. This approach is schematically presented in Fig. 1.3b. ZnO nanocrystals dispersed in methanol are brought up uniformly on the substrate by spin coating as seeds for the following nanorods aqueous growth process. Adhesion of the nanocrystals is improved by thermal treatment of the substrate. Implementation of such a seeded substrate leads to the growth of high-density ZnO nanopillars with preferential orientation (cf. Figs. 1.3b, 1.4).

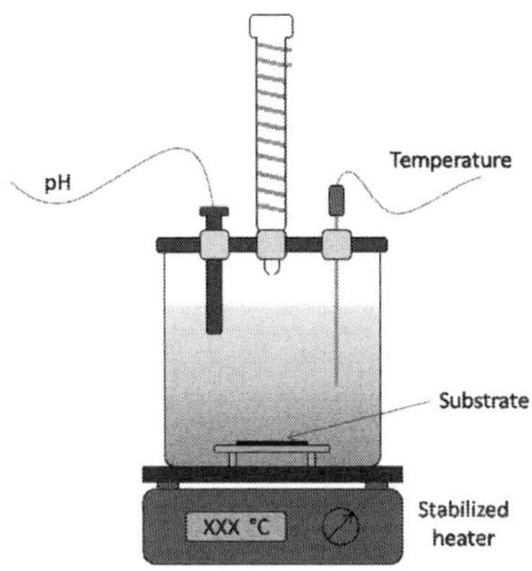

Figure 1.2 Schematic view of chemical bath deposition reactor.

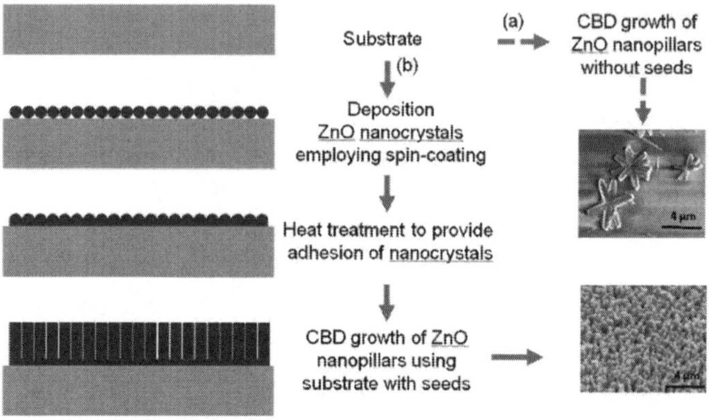

Figure 1.3 Schematic view of chemical bath deposition approach. Growth without seeds (a) leads to formation of low-density structures such as "flowers." Implementation of substrates with seeds (ZnO nanocrystals) for CBD growth leads to the formation of high-density ZnO nanorods with preferential orientation (b).

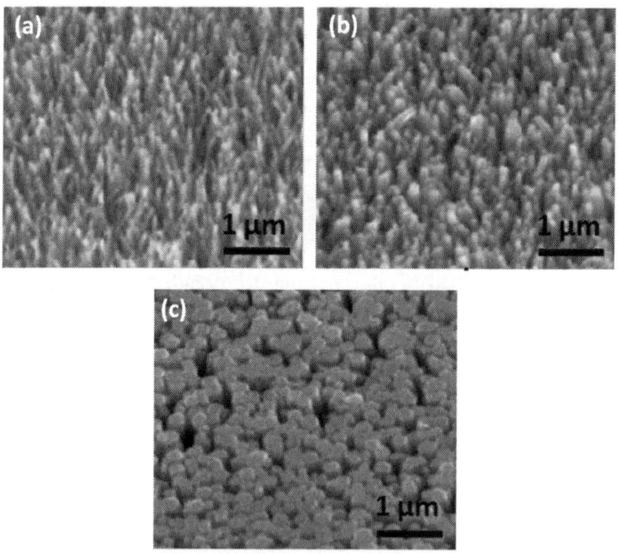

Figure 1.4 SEM images of ZnO nanopillars grown in 3 h on (001) silicon wafer at 90°C. Solution concentrations: (a) 0.01 mol/L; (b) 0.05 mol/L, (c) –0.075 mol/L.

The length and thickness of the nanopillars can be controlled by variation of solution concentration, temperature, and process duration [14, 15]. Figure 1.4 demonstrates influence of the solution concentration on the resulting nanopillars morphology. All other growth parameters were the same for all three samples.

Strong dependence of the thickness of the layers or of the length and diameter of the nanostructures on the process temperature and pH-values of the solution as well as on the concentrations of the source materials and their ratio in the solution not only provides the possibility to control the growth process but also means that properties of the grown structures can be influenced by the instability of these process parameters.

CBD growth of ZnO occurs on the surface of the substrate, which serves partly as the catalysator of this growth. However, this reaction can also occur in the solution leading to the formation of agglomerates of clusters and crystallites and resulting in the visible opacity of the solution. The agglomerates can also attach them to the substrate and influence the homogeneity and parameters of the deposited layers and nanostructures.

The polycrystalline underlayer formed by seeds (nanocrystals) schematically shown in the Fig. 1.3 and reported in literature [14, 16] provides a contact layer underneath of ZnO nanopillars. Such a layer can be of advantage if, for example, CBD growth is used to provide an upper contact layer and light outcoupling (nanopillars) from GaN-based LEDs [17].

In principle, CBD growth can be realized on different wafers—glass, FTO, Si, metal, etc. And this is an advantage of this method. However, for certain applications, it is an advantage to provide a selective growth of ZnO nanostructures—on the certain parts of the substrate. This can be realized for CBD approach by modifying the process growth parameters, e.g., reducing growth temperature and employing glass substrates with patterned silver layers. Homogeneous ZnO nanorod arrays have been grown on silver and no growth was observed on glass providing bottom-up generation of patterned nanopillars arrays [9].

One of the disadvantages of the CBD method is the low usage of the source material, which is typically below 10%, since due to agglomeration in the solution it is necessary to renew the solution after a certain process time. Therefore, it is also difficult to produce very long nanowires (20–30 μm) in one step without changing the solution. In order to fabricate such long nanowires, the solution should be changed several times leading to growth duration up to 20 h. Such long nanowires on FTO substrate are necessary, for instance, for photovoltaic applications such as dye-sensitized solar cells. A special two-step approach, including CBD growth of initial nanopillars with typical length of 2 to 3 μm and further rapid vapor phase growth of ZnO nanopillars, has been developed recently [18]. This approach, which provides high-density and high-quality arrays of nanopillars on FTO substrates with growth rates of 20 μm/h (about 10 times higher than the CBD growth rates on FTO substrates), is described in detail below.

1.4.2 Electrodeposition

Generally, the ZnO ED is carried out in a similar temperature-controlled chemical batch reactor at temperatures from 70 to 90°C and under ambient pressure [19]. The solution for ED consists of zinc nitrate hydrate ($Zn(NO_3)_2 \cdot 6H_2O$) and hexamethylenetetramine ($C_6H_{12}N_4$). The chemicals are solved in deionized water, resulting in a transparent solution [19]. The ED can be realized in two ways. In the first approach, ZnO nanorods are grown on substrates with conducting surface layer—for instance, silicon substrates covered with Ti/Au layer (20 nm Ti/200 nm Au), FTO or ITO substrates—at a cathodic potential of about 1 V. In this case, the Pt layer serves as a counter electrode (cf. Fig. 1.5). The distance between both electrodes is about 2 cm. Well-oriented ensembles of ZnO nanorods can be fabricated in such a way on different wafers, length and thickness of the nanopillars can be controlled by variation of the process parameters similar to CBD approach (cf. Fig. 1.6) [15, 19, 20].

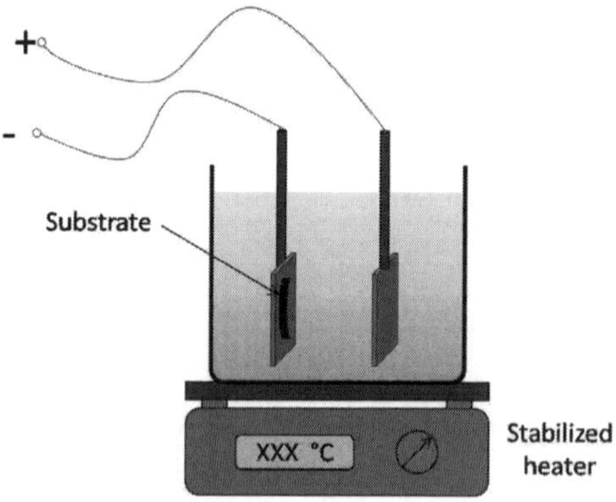

Figure 1.5 Schematic view of electrodeposition reactor with substrate (−) and Pt counter electrode (+).

Figure 1.6 SEM images of ZnO nanopillars grown on ITO wafer $c = 0.05$ M, $I = 1.5$ mA/cm^2 at 65°C (a) and 70°C (b).

An alternative ED approach to produce ZnO nanostructures is to grow nanopillars between two conductors deposited on the substrate and so to solve the problem of extremely complicated postgrowth contacting as well as fabricating self-organized nanopillars there where they are needed as the part of such nanodevices like transistors or sensors. Ti/Au stripes deposited on silicon substrate are implemented in this case. Such stripes acting as counterelectrodes can be realized employing, for instance, e-beam lithography (cf. Fig. 1.7a) [15, 20]. The fabricated interdigitated structures were 250–6000 nm wide and 400–1200 nm apart (cf. Fig. 1.7b). The bias voltage could be directly connected to both Ti/Au contacts and no external Pt electrode is necessary. In order to get control on the fabrication of separated nanowires between the stripes, the influence of the electric field was studied by varying the bias voltage between neighboring conducting metal stripes in the range from 0.5 to 2 V. High density of nanopillars on top of the stripes was obtained if the voltage was in the range 1–2 V (cf. Fig. 1.8a). The formation of ZnO nanostructures takes place on the cathode side (cf. Fig. 1.8b). Density of nanopillars is also influenced by solution concentrations. Increasing the voltage above 2 V leaded to layer-like growth instead of formation of nanowires.

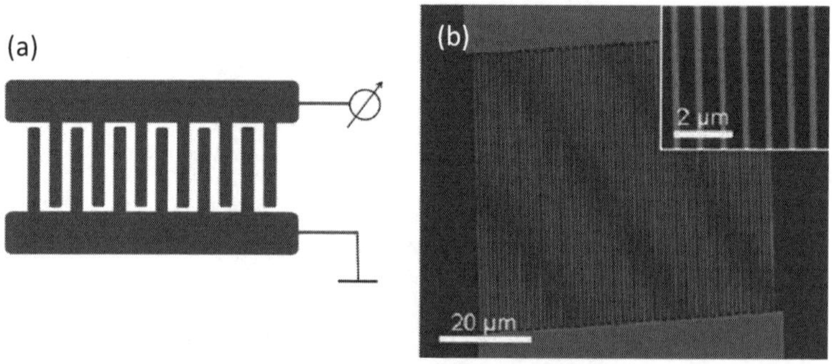

Figure 1.7 Schematic view (a) and SEM images (b) of a substrate with Ti/Au stripes for electrodeposition.

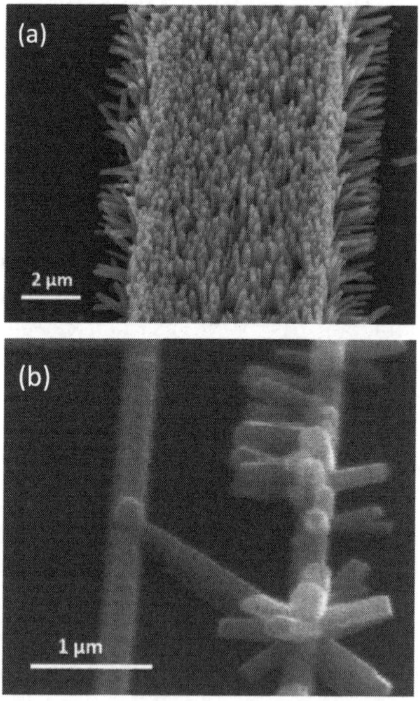

Figure 1.8 SEM images of ZnO nanorods grown on a substrate with Ti/Au stripes by electrodeposition at 70°C employing $U = 1.0$ V: high density of ZnO nanorods grown on a 6 µm-wide cathode (a) ZnO nanorod grown between two Ti/Au metal stripes and further nanorods grown on the 0.25 µm cathode stripe (b).

Nanorod density decreases drastically if the applied voltage is reduced from 1 to 0.5 V. The growth started predominantly at the edges of the metal stripes providing nanopillars between the stripes. In this case, of low-voltage deposition, the polarity of the applied voltage did not influence the growth process and no preferential growth on one of the contacts was observed. ZnO growth was not observed in absence of voltage between the stripes. Single nanowires between two stripes were obtained under the optimal growth conditions employing the voltage of about 0.5 V (cf. Fig. 1.9) [15]. These results demonstrate principal possibility of fabrication of self-contacted nanopillars employing ED providing partly self-organized nanodevice fabrication.

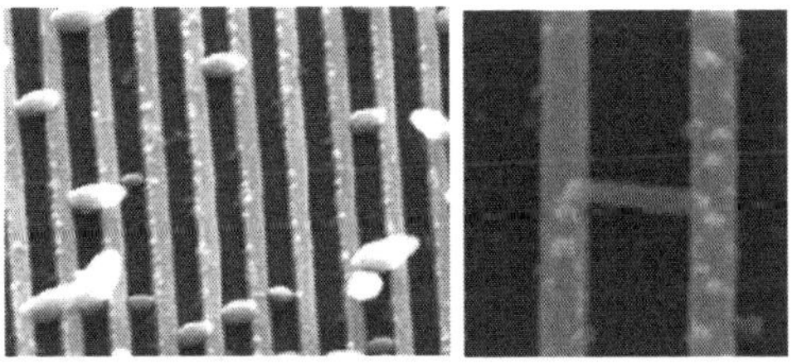

Figure 1.9 SEM images of ZnO nanopillars grown by electrodeposition at 70°C employing U = 0.5 V between two 0.25 μm Ti/Au metal stripes.

1.4.3 Successive Ionic Layer Adsorption and Reaction

Successive ionic layer adsorption and reaction is one more fabrication method for ZnO layers and nanostructures and also for other oxides, which is not really well known. The technique was firstly developed in 1985 [6]. This approach is presented schematically in Fig. 1.10 and can be described as follows. The source solutions are placed in separate vessels and the substrate is dipped in turn in the reactor with the cation precursor and then (also after rinsing the substrate with appropriate solvent, typically water or drying with N_2 flow) in the reactor with anion precursor. Thin films are grown layer by layer employing repetition of this process. This is to certain extent similar to ALD but from the liquid phase instead of the gas phase and at even lower temperatures (near to room temperature), so this method is also of special interest for depositing of very thin layers on nanostructures providing core–shell structures or solar cells with extremely thin absorber.

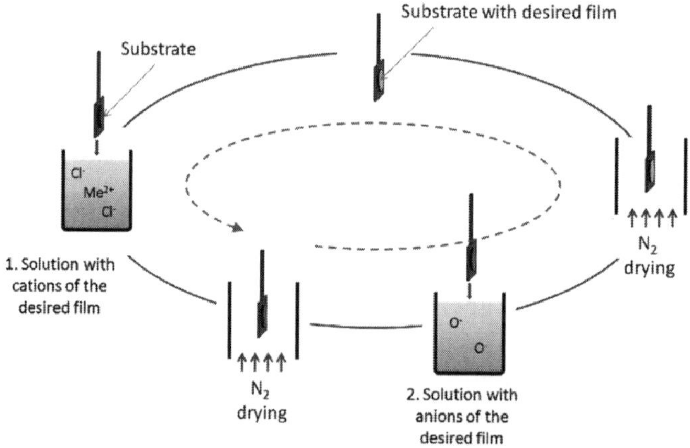

Figure 1.10 Schematic view of successive ionic layer adsorption and reaction approach.

ZnO SILAR growth was first employed for fabrication of films with thickness up to 100 nm. Successive immersion of glass substrate in a dilute solution of Zn^{2+}-ammonia complex at room temperature and in water heated to 96°C. Computer controlled electropneumatic SILAR deposition system was employed to provide reproducibly up to 35 emersion cycles. In the case of 35 cycles film thickness was about 100 nm, reducing number of cycles to 6 led to film thickness of 20 nm. As-grown films were of hexagonal (zincite) structure with a preferred *c*-axis orientation perpendicular to the surface of glass substrate [21]. Lupan et al. have demonstrated SILAR deposition of undoped and Sn, Ni-doped nanostructured ZnO thin films on glass substrates at room temperature. The average grain size of the deposited films was from 220 to 265 Å. The samples were employed for sensors fabrication. The variation in resistivity of the ZnO film sensors was obtained with doping and post-deposition rapid photothermal processing in vacuum and N_2 ambient. The nanostructured ZnO sensor showed higher ammonia sensitivity than that to NO_2. An enhanced NO_2 sensitivity was noticed with the ZnO films doped with 4 at.% Sn and higher NH_3 sensitivity was obtained by 4 at.% Ni doping of zinc oxide thin films [7].

Successive ion layer adsorption reaction approach is promising for fabrication of nanoheterostructures and we present implementation of SILAR approach for fabrication of ZnO-based core–shell nanostructures at the end of this chapter.

1.5 Gas-Phase Growth

1.5.1 Vapor Phase Transport and Chemical Vapor Transport

Physical vapor transport and CVT in general describe all growth methods that employ carrier gas for transporting precursors to the substrate. Compared with other growth methods, VPT is a relatively simple approach that uses pure elements as precursors, e.g., Zn and O_2 for the fabrication of ZnO avoiding impurities of the semiconductor resulting from byproducts of the precursor. Zn is transported to substrate employing inert carrying gas. Schematic view of the VPT system with two separated gas lines for the Zn-source and the oxygen gas is shown in Fig. 1.11 [22]. The VPT reactor is placed inside of a three-zone oven so, that the temperatures of the source material and the substrate can be varied separately. When an atom of the source material is transported via a carrier gas to the substrate it can be adsorbed and moves around on the substrate depending on its kinetic energy. Some atoms will pass over back to the gas phase when they gain energies higher than the adsorption energy due to local temperature fluctuations. It has to be ensured that the probability of this process is lower than of that atoms bound to the substrate. This occurs preferred at defects like dislocations since the atoms tend to stick to the positions with the lowest energy, thus positions with as much as possible bounding partners. Therefore, the substrate temperature has to be chosen very carefully.

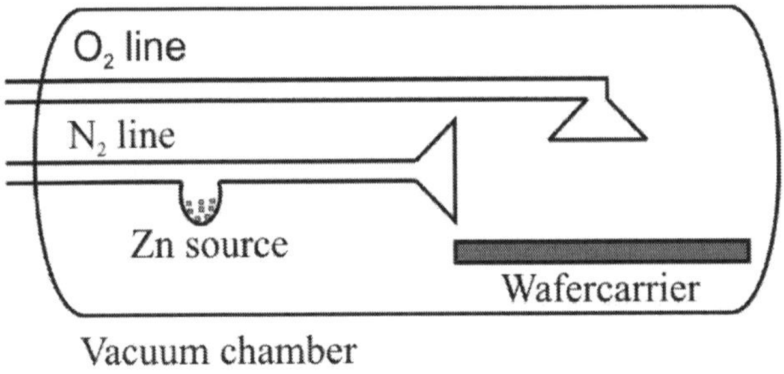

Figure 1.11 Schematic view of a vapor transport growth approach to ZnO nanowires fabrication.

In our CVT approach the high-purity Zn source material is heated to 470°C, evaporated and transported by N_2 as the carrier gas to the substrate. The substrate temperature is usually kept at 800–900°C and reactor pressure is typically maintained at 30 mbar. Furthermore, there is no need for any additional catalyst, the nanopillars grow self-organized on the substrate. The growth can be described using the Volmer–Weber model. The molecule arriving at the substrate form nucleus and if the growth rate is chosen properly these nuclei start to grow into nanopillars resulting in high crystal quality material due to small footprint on the substrate. SEM images of ZnO nanopillars grown on 6H-SiC-substrate are shown in Fig. 1.12. The nanopillars start to grow directly on the substrate without any additional nucleation layer giving indication of the Volmer–Weber growth model. ZnO nanopillars grown on GaN templates are shown in Fig. 1.13. Growth on GaN provides typically a much higher density of nanopillars compared to nanopillars ensembles grown on SiC due to the lower GaN/ZnO lattice mismatch of only 1.8%. More details of the VPT growth process can be found in [4, 20, 22, 23, 67].

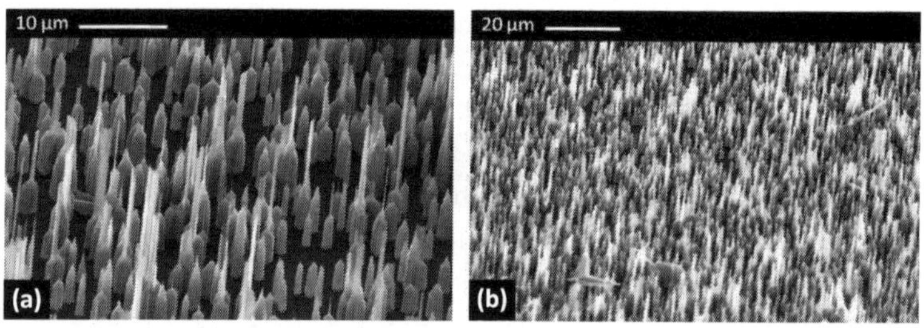

Figure 1.12 SEM images of ZnO nanowires grown on 6H-SiC substrate at a temperature of 800°C.

Figure 1.13 SEM images of ZnO nanowires grown a sapphire/GaN templates at a substrate temperature of 800°C.

Another approach to nanostructure fabrication is implementation of the catalyst assisted growth (typically vapor-liquid-solid growth) which was already invented in the 1960s for the fabrication of Si microstructures [24]. This requires a deposition of metal islands on the substrates before ZnO growth. An advantage of the catalytic growth is the possibility of controlling of the position, density and diameter of the nanopillars but the growth temperature has to be higher than the melting point of the catalyst. In the case of ZnO usually Au is used as a catalyst, thus the substrate temperature has to be above 1000°C. Furthermore, the catalyst might lead to a higher impurity concentration in the nanopillars. The process of a catalytic growth can be described as follows: Zn, which arrives at the substrate forms with the predeposited metal droplet an eutectic alloy. When this eutectic alloy saturates the growth of nanopillars starts. The metal catalyst remains on top of the nanopillar.

In [25] the synthesis of ZnO nanowires is described using a silicon substrate coated with 50 Å of Au. The diameter of the obtained nanowires was between 80 and 120 nm with the length from 10 to 20 μm. In this approach the substrate temperature was kept in the range of 900–925°C and Ar was introduced into the reactor as a carrier gas. XRD investigations of these samples indicate the high crystallinity of the nanowires.

In addition, Cu can act as a catalyst for ZnO nanowire fabrication. Li et al. reported successful growth of almost vertically oriented ZnO nanowires on Si at a substrate temperature from 850 to 950°C [26]. TEM investigations indicated single crystallinity of the wires, which also show strong room temperature PL emission centered at 325 nm.

In [27], the influence of different catalysts (Pt, Ag and Au) on the formation of nanowires was compared. Implementation of Au as a catalyst results in the thinnest and longest nanowires if high substrate temperatures of about 800°C are employed. The optimal substrate temperature for Ag as a catalyst is much lower, about 500°C, since Ag begins to oxidize at higher temperatures. It should be noticed that for a substrate temperature of 500°C the growth process is vapor-solid without a liquid state of the catalyst. Pt does not form droplets on the substrate but the layer cracks when heated above 500°C. These cracks serve as nucleation sites for Zn vapor absorption.

Long nanowires on FTO substrate are necessary, for instance, for photovoltaic applications like dye-sensitized solar cells. The disadvantage of chemical solution approaches is the relatively low growth rate of nanowires, which is limited by approximately 3 μm/h. A special two-step approach to fabrication of long ZnO nanopillars with the high growth rate on FTO substrate have been developed recently [18]. This approach includes CBD growth (described above) of initial nanopillars with typical length of about 2 μm and further rapid vapor transport growth of ZnO nanopillars. The growth temperature is limited in this case by approximately 600°C. Otherwise the properties of FTO degrade strongly. Under such low temperatures, no oriented vapor phase growth of ZnO nanopillars on FTO wafer was observed without the CBD nanowire seeding layer. This two-step approach provides high-density and high-quality arrays of nanopillars on FTO substrates with growth rates of 20 μm/h (cf. Fig. 1.14). This growth rate is about 10 times higher than the CBD growth rates on FTO substrates.

Figure 1.14 SEM images of ZnO nanowires grown by vapor transport at 600°C for 90 min on CBD grown "seeding" nanowires on FTO substrate.

1.5.2 Chemical Vapor Deposition

At present metal organic chemical vapor deposition (MOCVD) became the most used industrial approach to the fabrication of compound semiconductor. This method is widely used for production of LEDs. A great advantage of MOCVD over other growth methods is that the precursors, which are usually liquid or solid, are stored in stainless steel cylinders outside the reactor cabinet at almost room temperature. Furthermore, the requirements on the vacuum system are much lower compared to MBE since the reactor pressure ranges usually between several mbar up to atmospheric pressure. MOCVD can be successfully employed for the fabrication of ZnO nanostructures with well controllable diameter, length, and densities just by choosing the appropriate growth parameter. We have fabricated ZnO nanostructures in a vertical Thomas Swan 3 × 2″ system with a 3 plenum close coupled showerhead keeping the precursors separated from each other before they reach the substrate and so avoiding unwanted prereactions. Diethylzinc (DEZn) and laughing gas (N_2O) were used as the precursors for zinc and oxygen, respectively. The typical substrate temperature was varied between 800 and 1000°C at a reactor pressure between 100 and 400 mbar. The MOCVD growth process can be divided into several stages: transport of the molecules to the substrate, chemical reaction, and adsorption at the surface, and transport of the precursor byproducts from the substrate to the exhaust.

Figures 1.15–1.18 present the influence of different growth parameters on the formation of ZnO nanostructures under MOCVD growth. All nanostructures were grown self-organized without employing any catalyst on *a*-plane sapphire.

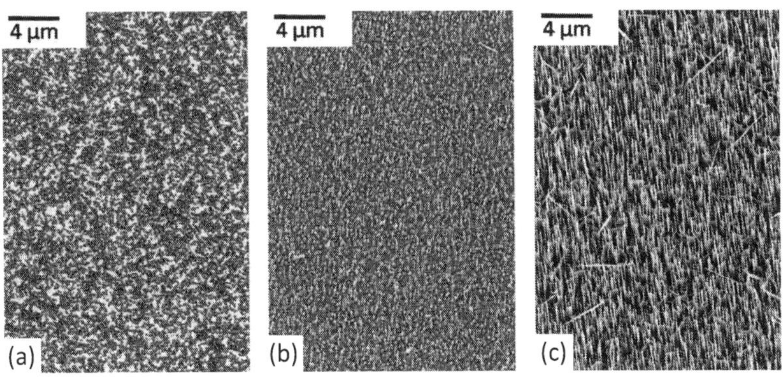

Figure 1.15 SEM images of ZnO nanopillars fabricated at different growth durations of 5 min (a), 30 min (b), and 60 min (c).

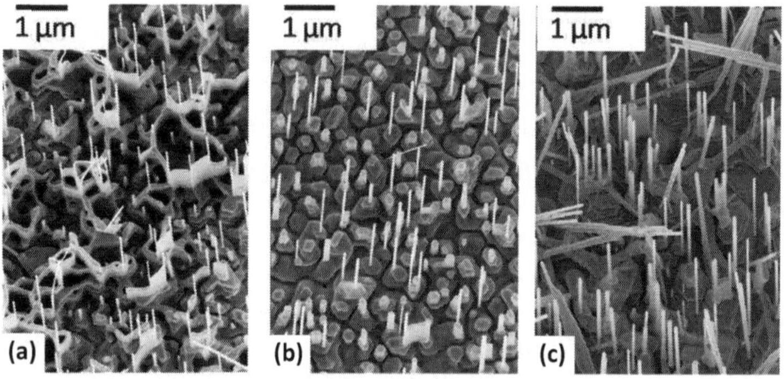

Figure 1.16 SEM images of ZnO nanopillars fabricated at substrate temperatures of 800°C (a), 900°C (b), and 1000°C (c).

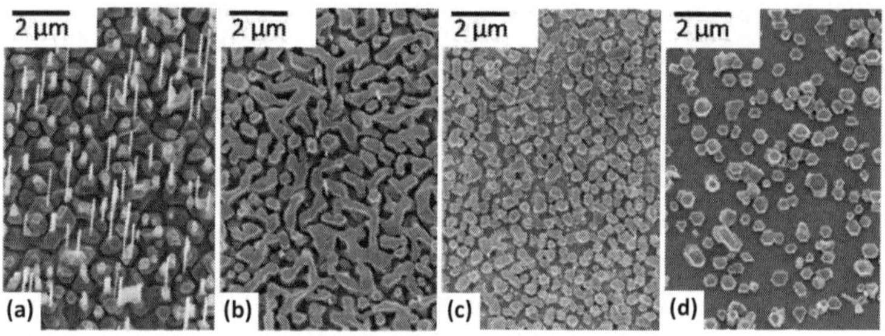

Figure 1.17 SEM images of ZnO nanopillars fabricated at VI/II-ratios of 220 (a), 700 (b), 3500 (c), and 7000 (d).

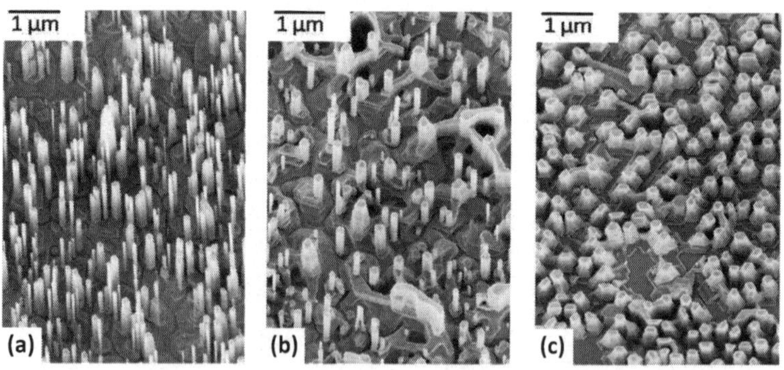

Figure 1.18 SEM images of ZnO nanopillars fabricated at different reactor pressures 200 mbar (a), 300 mbar (b), and 400 mbar (c).

Figure 1.15 shows the influence of the growth duration ranging from 5 to 60 min on the formation of nanopillars. The nanopillars do not start to grow directly on the substrate, but first three-dimensional islands grow self-organized and then occurs the 3D nanopillars growth. This effect is especially obvious for a relatively short growth time of about 5 min (cf. Fig. 1.15a). The growth of nanopillars begins when these islands start to coalesce since coalescence of the islands leads to compressive strain in the in-plain direction. If the underlying nucleation layer is doped with appropriate atoms, it can be used as a back contact on insulating substrate (for instance, sapphire). When the growth duration is increased, the density of the nanopillars remains almost constant at about 4.5×10^8 cm^{-2}. In contrast the average diameter increases linearly from 33 nm to 90 nm for a growth time of 15 min and 60 min, respectively (cf. Figs. 1.15b,c). Furthermore, the length of the nanopillars increases for the same durations from 800 to 1500 nm confirming that the preferred growth direction is along the *c*-axis.

One of the most important growth parameters is the substrate temperature; in general, the substrate temperature for the growth of nanopillars (pronounced 3D growth) is higher compared to the optimal growth temperature for layer fabrication. In Fig. 1.16, it can be seen that at the temperatures about 800°C mostly nanowalls are growing with only very few and thin nanopillars on top. When the temperature is increased, the formation of nanopillars is promoted. This can be explained by a higher mobility of the molecule for higher temperatures arriving at the substrate. The molecules aim to occupy the rough tip of the nanopillar, which provides positions with higher binding energy and not the side facet with lower surface energy. When the substrate temperature is low, the molecules do not gain enough energy to reach the tip leading to an increase growth rate of the side facets and the formation of nanowalls or layers at lower temperatures. However, certainly MOCVD growth also strongly depends on other parameters.

The diameter of the nanopillars can be controlled during the growth using appropriate VI/II ratios. In principle low VI/II ratios are leading to smaller diameters and high VI/II ratios to larger diameters. This behavior can be clearly observed in Fig. 1.17 demonstrating the formation of nanopillars with aspect ratios up to 20 only for the lowest VI/II ratio of 222. For higher VI/II ratios, the diameter is drastically increased, whereas the length is simultaneously decreased drastically. It should be pointed out that for higher VI/II ratios, the absolute amount of ZnO that is deposited on the substrate is decreased. This can be explained by the fact that there is much more reactive oxygen in the gas phase of the reactor leading to oxidation of the Zn atoms already in the gas-phase and also by the lack of Zn for higher growth rates.

Increasing the reactor pressure from 100 up to 400 mbar leads to the increase of both diameter and density of the nanopillars from 77 to 125 nm and from 8.5×10^7 to 8.8×10^8 cm^{-2}, respectively. In Fig. 1.18, SEM images of nanopillars grown at different reactor pressures are presented. It is clearly seen from the figure that for higher reactor pressures the nanopillars start to grow not on the nucleation layer but on nanowalls.

Rosina et al. [28] investigated the initial growth state of ZnO nanowires on *c*-plane sapphire at a substrate temperature ranging from 750 to 800°C employing a hot wall MOCVD reactor. Even after a very short growth time of only 1 min the formation of three-dimensional islands with nanorods on top was observed. The length and diameter of the nanorods were 350 nm and 20 nm, respectively. When the growth duration is increased it can be observed that the lateral growth rate is much smaller than the vertical growth rate resulting in high aspect ratios. The vertical growth rate saturates with the time. After a growth time of only 20 s the formation of a two-dimensional wetting layer was observed [29]. This can be seen as an indication of Stransky–Krastanov growth mode. ZnO nanorods revealed Zn polarity in contrast to the pyramids, which usually showed O-polarity. Hence, for the growth of nanorods nuclei with Zn-polarity are required.

Zhang et al. [30] reported the growth of ZnO nanowires using O_2 and DEZn as the precursors for oxygen and Zn, respectively. The use of pure O_2 as a source for oxygen enables fabrication of nanowires at relatively low growth temperatures of 475°C. Implementation of N_2O as an oxygen source requires substrate temperatures above 500°C otherwise the precursor will not be cracked.

Baxter et al. investigated the influence of the substrate on the formation of nanowires [31]. The MOCVD process was performed using $Zn(acac)_2$ hydrate and a mixture of oxygen and H_2O as the precursors. The substrate temperature was kept at 550°C. The growth parameters were kept constant for all experiments. As a result random oriented nanowires were obtained on $F:SnO_2$ and glass, whereas the growth of well oriented nanowires was observed on *a*-plane as well as on *c*-plane sapphire. Polycrystalline films were obtained on *c*-plane ZnO and Si substrates.

1.6 Pulsed Laser Deposition

Pulsed laser deposition (PLD) is at present industrially used growth technology, which is well developed for ZnO epitaxial layers and nanostructures growth. The process is realized in stainless steel chamber under background pressure up to 100 mbar as described below. Elemental or multicomponent materials can be evaporated by laser beam from the target in the vacuum chamber. So called plume of gas-phase material is created in this way and it reacts with the ambient provided by introducing gases through the gas inlet and condenses on the substrate placed on the heated holder. In the case of ZnO fabrication, the process can be carried out in oxygen ambient. Manipulator with several targets can be introduced into the system providing the possibility to manufacture doped materials, multilayer structures of different materials or superlattices. Typically, excimer lasers with wavelength of 193 nm or 248 nm as well as Nd.Yag lasers with wavelength of 355 nm are employed for the evaporation of the target. The typical frequency of the pulses is up to 50 Hz. A lens is employed in order to focus the laser beam on the target with the spot size of about 0.1 mm^2 and the angle of incidence typically 45°. In the PLD approach, the composition of the grown layers is typically directly defined by the target composition although it may deviate from the target composition due to different sticking coefficients of the ablated species at the heated substrate. Pressing of high-purity ZnO powder with the consequent sintering at the temperature of about 1100°C are employed for fabrication of the target. Dopants (for instance, P_2O_5 for doping with P) are mixed into the ZnO powder in order to fabricate doped ZnO.

Alternatively, poly- or monocrystalline wafers can also serve as the target, but such targets are much more expensive then the pressed ones. Implementation of pure Zn as the target is also possible if the process is carried out under oxygen environment. PLD approach can be further improved by employing shutters between the target and the substrate, making possible to pre-evaporate the target before the deposition process starts. While implementing PLD approach, it is important to avoid droplet formation on the growth surface, which can be caused by the ablation of such droplets from the target. It can be done by reducing laser fluence and implementing shorter wavelength laser radiation causing at the same time a decrease of the deposition rate. Nucleation and growth of ZnO as well as defect densities, electrical and optical properties are strongly influenced by the oxygen pressure [32].

PLD of ZnO nanostructures on different substrates have been demonstrated [16, 33–35]. A dense array of regular nanorods with a preferred orientation perpendicular to the substrate plane was obtained on Si and c-Al_2O_3, by PLD, without the use of a catalyst. Structures grown on Si had lower maximal intensity and smaller FWHM than those grown on c-Al_2O_3. Al diffusing into the ZnO from the substrate was supposed for the case of ZnO nanostructures growth on Al_2O_3 [33]. Lorentz et al. employed high-pressure PLD (background pressure about 100 mbar) for the formation of self-organized ZnO nanopillars on Al_2O_3, Si, SiC substrates and low-pressure PLD (background pressure below 0.1 mbar) for manufacturing of shell around nanorods or nucleation film on substrate [16, 34, 35]. Nanostructures of different length and diameter can be realized by varying parameters of the pulsed laser deposition.

1.7 ZnO Nanopowder-Based Technologies

As an alternative to the traditional growth methods from the gas and liquid phases described above, there is another group of highly cost-effective approaches to the fabrication of nanostructures-based devices employing nanopowders to fabricate nanocrystalline or nanoparticulate layers. Such layers are of great interest for sensing and photovoltaic applications. The most known approaches to deposit nanoparticulate layer are spin-coating and droplet-coating employing colloidal solution. To stabilize colloidal solutions and to avoid the formation of aggregates organic additives are added to the solution. After deposition of the film a firing step at elevated temperatures is needed to remove the organic additives. We have employed droplet coating for deposition of ZnO nanopowder delivered by Grillo Zinkoxid GmbH [20, 36]. The ZnO grain size was varied in several ranges: 5, 25, and 60 nm. Spin-coating and drying were followed by pressing of the layer in order to reduce porosity, to increase homogeneity of the layers and to improve surface planarity and attachment to the substrate. Some examples of such ZnO nanopowder layers are presented in Figs. 1.19–1.21.

Figure 1.19 FeSEM images of glass substrate as deposited droplet-coated with 25 nm ZnO nanopowder (a) and the sample after compression (250°C, 5 min, 40 MPa) (b).

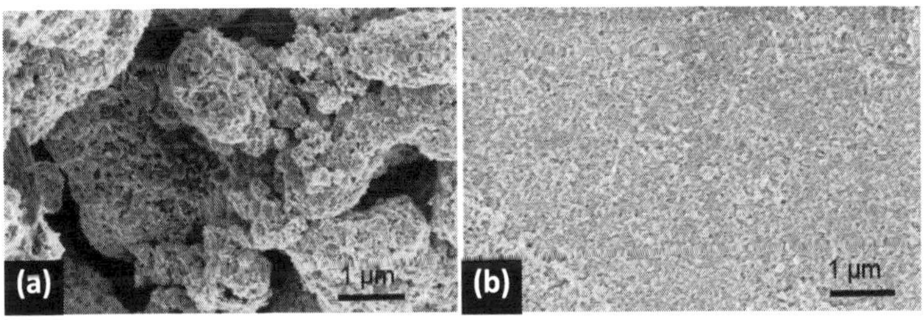

Figure 1.20 FeSEM images of glass substrate as deposited "droplet-coated" with the mixture of 25 nm + 5 nm ZnO Nanopowders (a) and the sample after compression (250°C, 5 min, 40 MPa).

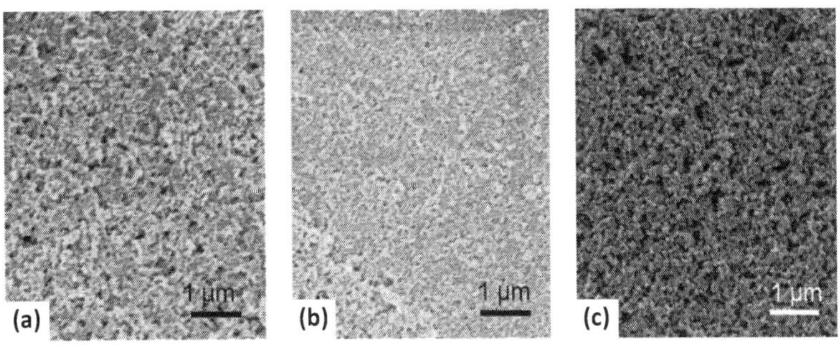

Figure 1.21 FeSEM images of glass substrate "droplet-coated" with different ZnO Nanopowders and then compressed at 250°C, 5 min, 40 MPa: (a) 25 nm ZnO nanopowder, (b) mixture of 25 nm + 5 nm ZnO nanopowders, (c) 60 nm ZnO nanopowder.

One more simple approach to deposit nanoparticulate layers is deposition of a paste by doctor blading. Lindström et al. introduced a new method for the fabrication of nanostructured electrodes on glass substrates for the use in electrochemical cells [37, 38]. They applied a suspension of TiO_2 nanoparticles in ethanol on a glass substrate and subjected the sample between two steel plates to pressure. Through the successful fabrication of dye-sensitized solar cells, it was shown that nanoparticles had a good electrical contact between each other and therefore the firing steps became redundant.

In our experiments, we apply a modified doctor blade method for the fabrication of thin nanoparticulate layers for sensing and photovoltaic applications. First, a seeding layer (identical to that used for seeded CBD growth described above) was formed on the FTO substrate and then nanoparticles dispersed in isopropyl alcohol have been deposited employing doctor blading. Then the samples were compressed and additionally they were annealed during the compression to 120°C to ensure good electrical contacts between the particles. The temperature is low enough for the process to be compatible even to plastic wafers.

Field emission scanning electron microscopy image of the as deposited ZnO nanoparticulate layer is presented in Fig. 1.19a and 1.20a for 25 nm nanopowder and for the mixture of 25 nm and 5 nm nanopowders, respectively. It is obvious that ZnO nanoparticles form agglomerates with the mean size of several micrometers, which stick loosely together. After the application of the hydrostatic compression, the agglomerates disappear and the nanoparticles are distributed homogeneously forming smoother layer (cf. Figs. 1.19b and 1.20b. The resulting layer became mechanically much more stable and sticking to substrate. The mean value of the porosity decreases with the decrease of the nanoparticles size (compare Figs. 1.21a,c for 25 nm and 60 nm particles) and can be further improved by mixing 20–30 nm powder with 5 nm powder (cf. Fig. 1.21b). The mean values of the porosity decrease with increasing process pressure. Sensors and DSSCs were fabricated on the base of such ZnO nanoparticulated layers giving evidence of good electrical contact between particles and providing possibilities for device optimization [36, 65]. The described ZnO nanopowder approach was successfully employed for fabrication of DSSCs showing ~5% efficiency under 10% AM 1.5 (10 mW/cm^2) [65].

1.8 Fabrication of Complicated ZnO-Based Heteronanostructures

1.8.1 Fabrication of Quantum Wells Embedded into ZnO Nanostructures

Several approaches have been successfully employed for fabrication of QWs embedded into ZnO nanopillars.

Molecular beam epitaxy approach employing oxygen-plasma cell as oxygen source was implemented for the growth of ZnO/ZnMgO single- and multiple QWs embedded into ZnO nanopillars [39, 40, 68]. Zn and Mg metal sources were evaporated in double zone effusion cells. Oxygen radicals were introduced with an oxygen RF plasma cell operated at 200 to 400 W. Wafers with ZnO nanopillars obtained by CBD or vapor transport described above were used as epiwafers for consequent $Zn_{1-x}Mg_xO/ZnO/Zn_{1-x}Mg_xO$ ($0 < x < 0.2$) QW structures fabrication. In situ reflection high-energy electron diffraction (RHEED) oscillations were used for the optimization of the heterostructures growth. The wafers with nanorods were introduced into MBE chamber and first heated at 800°C for 40 min without oxygen flow and later for 30 min in an oxygen plasma overpressure of 3.1×10^{-5} Torr. Then the first $Zn_{1-x}Mg_xO$ barrier of 70 nm thicknesses was epitaxially grown on the tips of the ZnO nanorods at 450°C, followed by the multiple quantum wells (MQWs) consisting of 10 period of 1.0–4.0 nm ZnO/7.0 nm $Zn_{1-x}Mg_xO$ and the structure was then finished with the upper 50 nm $Zn_{1-x}Mg_xO$ barrier. PL measurements on the samples obtained showed that the nanopillar QWs show bright PL, even at room temperature. ZnO peaks for different samples were observed at the same position, while shift of ZnO–QW and ZnMgO peaks was observed if well width or Mg composition were varied. Blue-shift of the QW peak position with respect to ZnO nanorod buffer have been observed due to the quantum confinement of electrons and holes.

High-quality ZnMgO/ZnO QW nanostructures were also grown employing PLD [41]. High-pressure PLD process at background pressure of about 100 mbar is more flexible concerning nanopillars doping, density and aspect ratio. This approach was employed for fabrication of ZnO/ZnMgO QW structures [42] and fabrication of ZnO indicating p-type conductivity [43, 44]. Further details on ZnO PLD can be found in recent reviews [45, 46].

Also MOCVD was used for fabrication of ZnO/ZnMgO QWs embedded into ZnO nanopillars [47]. Low-pressure metal-organic vapor phase epitaxy have been employed for manufacturing of nanorod quantum structures. The structures obtained were of uniform width and length. Typical diameters and lengths of obtained nanorods were in the range of 20–70 nm and 0.5–2 μm correspondingly and ZnO/ZnMgO superlattices with respective periodic thicknesses of 4.6–14.5 nm have been fabricated [47]. Quantum confinement of carriers in the nanopillars QWs was observed using photoluminescence spectroscopy.

Fabrication of such advanced QW nanostructures is an important step toward realization of nanostructures based LEDs and other nanodevices. The main challenge at present is still lack of the reproducible p-type doping of ZnO. That is why many groups at present try to implement other p-type materials to realize p–n junctions with ZnO as the n-side and employing, for instance, p-GaN, p-SiC, p-Si, p-Cu_2O, etc. as the p-side or alternatively growing ZnO nanopillars on GaAs wafer and so realizing As diffusion for supposingly p-type doping of ZnO nanopillars.

1.8.2 Combination of ZnO Nanostructures with Other Semiconductors

Until now, ZnO suffers from the lack of the fabrication of reproducible and long-term stable p-type material. Hence, there is a wide variety of different materials that may substitute p-ZnO to fabricate LEDs or further devices. Among the candidates are GaN, SiC, Si, and Cu_2O.

GaN seems to be very promising because of quite similar to ZnO physical properties and a well-developed LED fabrication technology. The emission spectrum from n-ZnO/p-GaN heterostructures varies from yellow to blue/UV and is demonstrated with a voltage either in forward or reverse direction [48–56]. The band diagram of a ZnO/GaN heterostructure using the Anderson model is shown in Fig. 1.22 with a valence and conduction band discontinuity of 0.12 eV and 0.15 eV, respectively [53]. A SEM image of ZnO nanostructures fabricated by VPT method on a MOCVD grown GaN template is shown in Fig. 1.23. The top layer of the template consists of p-doped GaN with a carrier concentration of 2×10^{19} cm^{-3}, which is separated from the underlying GaN buffer by a 200 nm thick AlGaN barrier to avoid recombination from the p-GaN with the intrinsic n-GaN. During the ZnO growth a Ga_2O_3 interlayer is formed providing deep nonradiative recombination centers; therefore a relatively low growth temperature is preferable. The room temperature cathodoluminescence spectrum shows a band edge emission at 3.23 eV and a broad defect related emission at 2.3 eV. Electroluminescence measurements for bias voltages ranging from 10 V to 25 V in forward direction are shown in Fig. 1.24. The EL as well as the CL spectra are dominated by a broad emission around 2.3 eV. Figure 1.24 shows a photograph of the sample with a bias voltage of 12 V generating green emission.

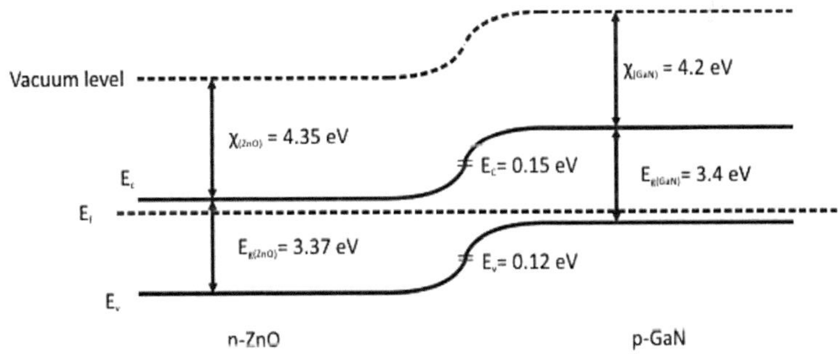

Figure 1.22 Band diagram of n-ZnO/p-GaN heterojunction using the Anderson model.

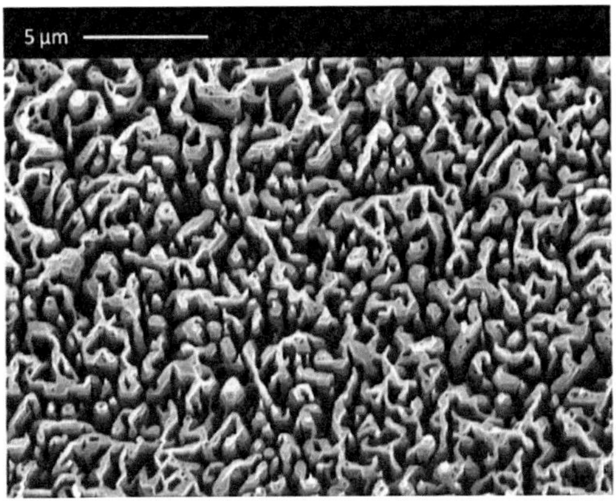

Figure 1.23 SEM image of n-ZnO nanostructures grown on p-type GaN templates using VPT growth method.

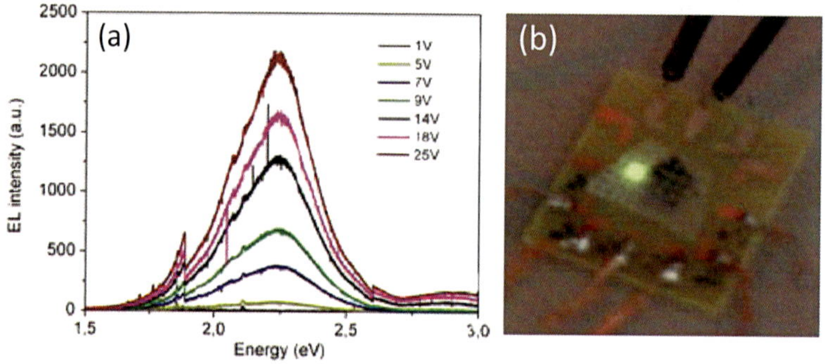

Figure 1.24 Room temperature EL spectrum of the n-ZnO/p-GaN hybrid LED for different bias voltages in forward direction (a) and a photo of the LED biased with a voltage of 12 V.

Manufacturing of a high brightness hybrid LED consisting of a p-type Mg-doped GaN layer grown via a CVD process on sapphire substrate followed by a vapor-solid fabrication of ZnO nanowires have been reported by Zhang et al. [49]. These structures show bright electroluminescence in UV-blue spectral region when the bias voltage is supplied in forward direction. A blue shift was observed with increasing bias voltage due to the modification of the band profile when an external voltage is supplied.

There are even reports on white luminescence from ZnO/GaN hybrid LED [51]. The sample consists of ZnO fabricated by pulsed laser deposition on p-doped GaN with a high density of structural defects at the interface. The white light results from a superposition of a blue emission at 440 nm attributed to a shallow donor—valance band transition with a yellow emission at 550 nm resulting from lattice defects in GaN. A great advantage of this approach compared to other white light LEDs is that there is no need for phosphors.

At the same time if ZnO is highly n-type doped it can be used as a transparent conductive oxide enhancing the current spreading and light outcoupling compared to conventional metal contacts [57].

In the case of CBD growth ZnO polycrystalline underlayer formed by seeds (nanocrystals) provides a contact layer underneath of ZnO nanopillars. Such a structure provides TCO layer and at the same time light outcoupling (nanopillars) from GaN-based LEDs [17]. Furthermore, the growth temperature especially for wet chemical growth processes are much lower compared to MOCVD growth temperatures resulting in much less diffusion of In in Incan QWs and thus causing no QW degradation.

GaN on ZnO

The lattice constant as well as the bandgap of GaN and ZnO are quite similar, therefore the growth of GaN on ZnO-substrates would result in a decreased dislocation density compared to the usually employed substrates such as Al_2O_3 and SiC with a relatively high lattice mismatch of 18% and 1.8%, respectively. Unfortunately ZnO decomposes in the environment which is required for the fabrication of high-quality GaN namely temperatures above 1000°C and hydrogen atmosphere. Figure 1.25 presents optical and corresponding SEM images of MOCVD grown GaN on ZnO bulk-substrate revealing partly decomposition of a quarter of a 2″ ZnO wafer. In the areas where the wafer still exists, the GaN layer is peeling off and the morphology is quite worse. When nitrogen is used as the carrier gas, the ZnO substrate does not dissolve but in photoluminescence measurements a response from neither ZnO nor GaN could be detected. Thus, a covering layer has to be employed to prevent the substrate from etching during the growth process, which is of special interest for nanostructures where the surface to volume ratio is much larger. This covering layer can be realized by a two-step process where the growth starts with a low temperature covering layer employing nitrogen as a carrier gas and with pulsed precursor flows followed by the high temperature growth with hydrogen as the carrier gas and pulsed precursor flows as well. Figure 1.26 shows the comparison of GaN layers grown with a continuous flow and a pulsed flow of the precursors indicating a drastic decrease of the surface roughness for the pulsed growth mode.

Furthermore, ZnO nanostructures can be used as a template for GaN nanostructures, which are usually fabricated on time and cost consuming prepatterned substrates. In contrast, ZnO nanostructures can be fabricated self-organized relatively easy with well controllable diameter, density and height just by controlling the growth parameter. In Fig. 1.27 the SEM picture of GaN grown around ZnO nanopillars are shown with some pyramidal shapes structures on top of the ZnO pillars. Low-temperature cathodoluminescence investigations provided evidence of the successful growth of GaN on ZnO nanowires. The ZnO band-edge emission could not be detected for low acceleration voltages but only for acceleration voltages above 15 kV. The Cl spectrum is dominated by broad emission bands between 1.5 and 3.3 eV due to structural defects.

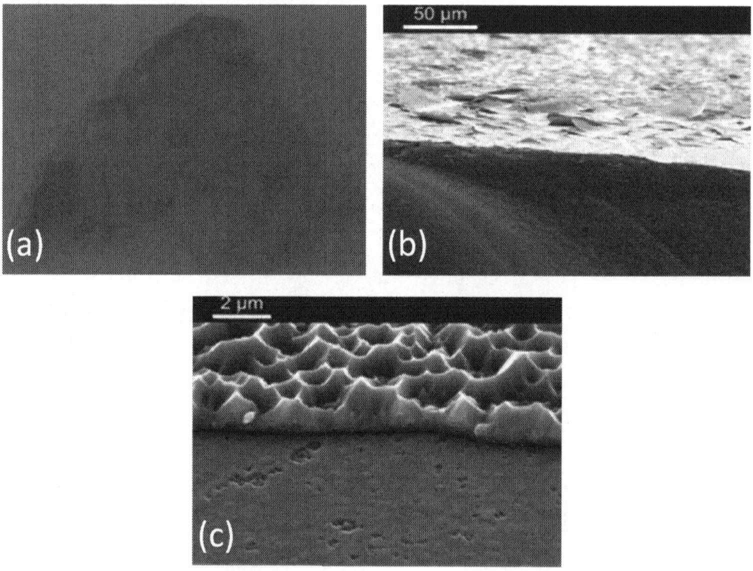

Figure 1.25 Optical (a) and SEM (b, c) images of GaN grown on ZnO bulk-substrate using MOCVD.

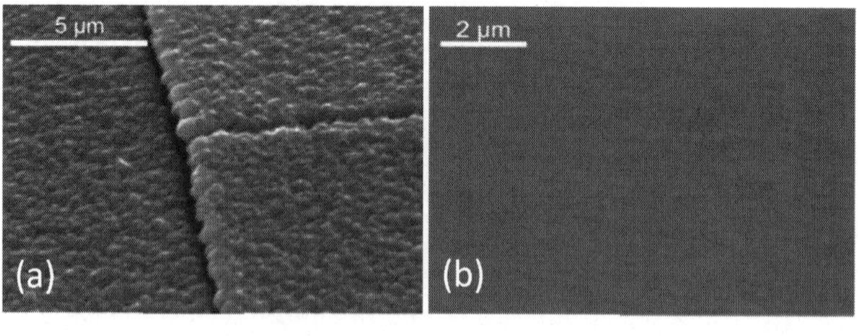

Figure 1.26 SEM images of GaN layers grown on ZnO templates using MOCVD continuous flow (left) and pulsed flow (right) growth process.

Figure 1.27 SEM image of GaN grown by MOCVD on VPT grown ZnO nanowires.

Fikry et al. have reported on the successful growth of GaN and even Incan QWs around ZnO nanostructures [58]. GaN cover layer was grown at 550°C in N_2 atmosphere in order to protect ZnO from decomposing during high temperature GaN growth at 1050°C in H_2 atmosphere. Position control of the nanostructures was achieved by the growth selectivity of ZnO on different planes on GaN pyramids.

Incan layers with different In content were fabricated also employing a low-temperature cover layer fabricated at 530°C followed by a high temperature growth at 680°C to 720°C resulting in In concentrations between 17% and 27% [59]. All samples showed a broad Incan related emission for the room temperature PL spectrum with a varying peak position due to the varying in content.

GaN overgrowth of nanostructures is very complicated and optimization of this process was made first employing layers growth. In [60] freestanding GaN layers were fabricated directly on Zn-polar ZnO substrates. The layers were grown at a substrate temperature of 850°C followed by a gas-phase etching process at 1000°C to remove the ZnO substrate. This approach can be further transferred to fabrication of GaN-based nanostructures on ZnO nanopillars. Obtained GaN nanostructures then can be attached to the required substrate and ZnO can be removed. This approach after certain modifications could be employed for GaN-based nano-LEDs fabrication.

ZnO-based core–shell nanostructures

Other promising p–n heterostructures are n-ZnO/p-Cu_2O. Implementation of such structures was tried for light emitting applications [61] and is of special interest for fabrication of sustainable electronics and solar cells. Different approaches (mostly MOCVD or sputtering) have been used for fabrication of plain Cu_2O layers [70, 71]. Implementation of SILAR approach described above for the growth of Cu_2O on a plain substrate led to fabrication of both layers and even semitransparent nanostructured flexible material consisting of Cu_2O nanowires shown in Fig. 1.28a. XRD measurements for this sample have shown the presence of only Cu_2O phase without indication of CuO (cf. Fig. 1.29). Until now, such nanowire layers were never reported for SILAR grown Cu_2O. At the next step, CBD-grown ZnO nanopillars were employed as the substrate for Cu_2O SILAR growth at the same conditions as on the glass wafer leading to fabrication of a core–shell nanostructure presented in Figs. 1.28b,c. These results give an indication that further development of SILAR technology could lead to cost-effective fabrication of advanced p–n structures [62].

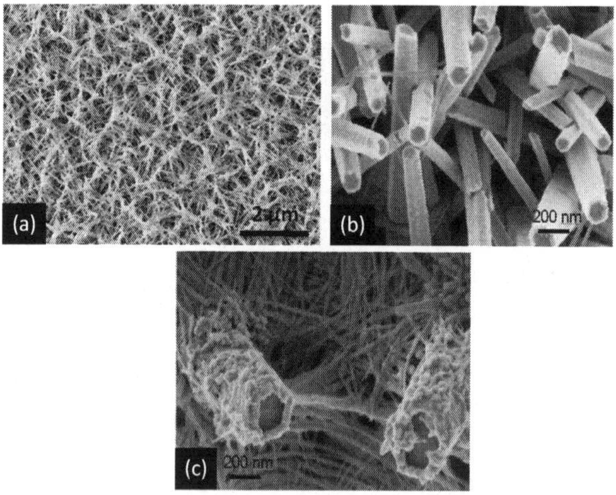

Figure 1.28 FE-SEM images of: SILAR-grown Cu_2O nanowires on a glass substrate (a) and under the same conditions SILAR-grown core–shell nanostructures on ZnO nanorods (b); initial deposition of thin shell and further growth of thicker shell and formation of nanowires between nanopillars (c); ZnO nanopillar is covered from the sides while the (0001) upper plane is free of deposition.

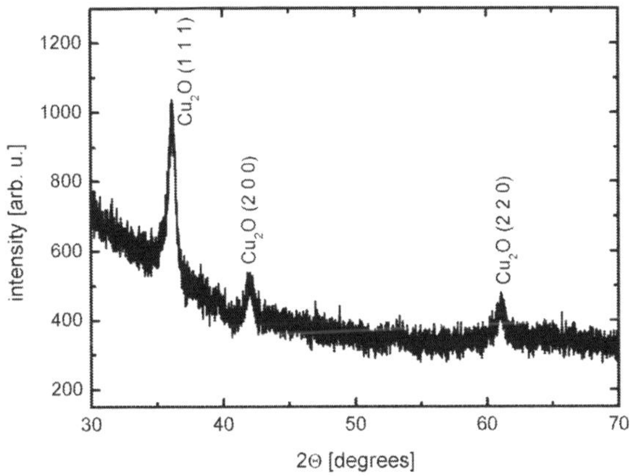

Figure 1.29 2Θ/ω scan for SILAR-grown Cu_2O nanowires on a glass substrate (shown in Fig. 1.28a).

Cu_2O/ZnO core–shell structures grown by wet chemical deposition (dip-coating of wafers with ZnO-nanorods) were recently demonstrated [72]. Initial ZnO nanorod arrays were grown by CBD on Si substrates. Cu_2O nanoparticles with a diameter of about 5 nm were forming the shell with the thickness of about 10 nm around ZnO nanorods. The obtained core–shell structures demonstrated an improved photoresponse with ultrafast and extended high sensitivity from UV to the visible range.

It is also worthwhile to mention ZnO based core shell structures for photovoltaic applications like ZnO/CdS. Fabrication of ZnO/CdS core/shell nanorod arrays by a two-step method have been reported recently [63]. Initial ZnO nanorod arrays were grown by ED on FTO substrates. At the next step SILAR approach was employed to grow CdS nanocrystalline film on the ZnO nanorods providing formation of a core–shell nanostructures. CdS increases absorption in the solar spectrum region. Obtained ZnO/CdS core–shell structures were implemented for fabrication of DSSCs and their performance was compared to that of ZnO nanopillars based DSSCs revealing 13-fold enhancement in photoactivity [63].

1.9 Summary

Fabrication of ZnO nanostructures, ZnMgO/ZnO QWs as well as ZnO/p-semiconductor using MOCVD, vapor or chemical phase transport, pulsed laser deposition, MBE, CBD, ED, SILAR have been discussed. Comparison of typical parameters for different types of ZnO nanorods fabrication methods is presented in the Table 1.1. The density, length and diameter of ZnO nanopillars are well controlled by changing the critical growth parameters for each method. PVT, MOCVD approaches provide the widest family of nanostructures. Chemical solution growth methods are the most cost effective and simple. Nanostructures grown by VPT, MBE and PLD have the lower impurity doping levels than the structures fabricated by MOCVD, CBD, ED, and SILAR. Wet chemical approaches as well as VPT and sputtering can provide fabrication of nanostructures under low enough temperatures thus giving the option of implementation of flexible substrates like plastic ones.

Table 1.1 Comparison of typical parameters for different ZnO nanorods fabrication methods

Method	Growth rate (μm/h)	Substrate temperature (°C)	Structural quality	Reactor pressure (hPa)	Costs
CVD	1–1.5	600–1000	++	100–500	+/-
CVT	~1	800–900	++/+	30–500	+
VPT	~1	800–900	++	30–500	+
VPT*	~20	350–600	++	30–500	+
PLD	2–10	500–800	++/+	5–30	+
ED	2–8	70–90	+/-	10^3	++
CBD**	0.2–0.7	60–90	+/-	10^3	++
Sputtering***	~1	400–700	+/-	10^{-3}–1	+

* Low-temperature VPT employing nanostructures grown by other methods as a substrate.

** Growth stops after ~2–3 h due to agglomeration in the chemical solution; the solution must be renewed.

*** Typically used for overgrowth on nanostructures grown by other methods (for instance, of ZnMgO on ZnO).

The possibility to employ several fabrication techniques is of special importance for the realization of unique device structures. ZnO nanowire arrays, prepared by metal organic chemical vapor deposition (MOCVD), were used as seed layers for fabricating large-area well-aligned ZnMgO nanowire arrays, using radio frequency (RF) magnetron co-sputtering [64]. Also, well-developed nanopillar fabrication technique like VPT was combined with MBE technology for ZnMgO/ZnO multi QW fabrication. Embedding of ZnMgO-ZnO QW structures into ZnO nanopillars have been demonstrated employing MOCVD, MBE and PLD presenting significant progress toward nanodevices realization. Such nanostructures are promising as the basis for future nano-LED structures, providing a potential of up to 10^{10} devices/cm^2. They can even be combined with glass or plastic substrates and the nanopillars based QWs should still have a high efficiency due to their small footprint on a susbstrate resulting in a low defect density. First results of the device applications of ZnO nanostructures are presented. Implementation of nanopowders is also a promising direction of ZnO nanotechnology development.

Acknowledgements

The authors would like to thank M. Karsten, K.-H. Lachmund, A. Scmidt, and D. Rümmler for their technical assistance and D. Rümmler for FeSEM investigations.

Samples manufacturing, analysis, and discussions by Dr. B. Postels, Dr. E. Schlenker, Dr. A.-C.Mofor, and Dr. A.-H. El-Shaer during their PhD studies are gratefully acknowledged.

This work was supported by Deutsche Forschungsgemeinschaft (DFG), EU (project NANDOS) and BMBF (project Nano-Quit). Special acknowledgements to the company Grillo Zinkoxid GmbH for providing special ZnO nanopowders, and we would like to thank Dr. K. Diekstall and Dr. B. Jahn from Grillo Zinkoxid GmbH for fruitful discussions, support, and enthusiastic approach.

References

1. Bagnall, D.M., Chen, Y.F., Zhu, Z., Yao, T., Koyama, S., Shen, M.Y., Goto, T. (1997). Optically pumped lasing of ZnO at room temperature, *Appl. Phys. Lett.*, **70** (17), pp. 2230–2232.
2. Kawasaki, M., Ohtomo, A., Ohkubo, I., Koinuma, H., Tang, Z.K., Yu, P., Wong, G.K.L., Zhang, B.P., Segawa, Y. (1998). Excitonic ultraviolet laser emission at room temperature from naturally made cavity in ZnO nanocrystal thin films, *Mater. Sci. Eng. B*, **56** (2–3), pp. 239–245.

3. Look, D.C. (2001). Recent advances in ZnO materials and devices, *Mater. Sci. Eng. B-Solid*, **80** (1–3), pp. 383–387.
4. Bakin, A., Waag, A. (2011). ZnO Epitaxial Growth. Invited chapter for *Comprehensive Semiconductor Science and Technology* Encyclopaedia ELSEVIER, edited by R. Fornari ISBN: 978-0-444-53143-8.
5. Sankapal, B.R., Sartale, S.D., Lokhande, C.D., Ennaoui. A. (2004). Chemical synthesis of Cd-free wide band gap materials for solar cells. *Solar Energy Mater. Solar Cells*, **83**, pp. 447–458.
6. Nicolau, Y.F. (1985). Solution deposition of thin solid compound films by a successive ionic-layer adsorption and reaction process, *Appl. Surf. Sci.*, **22–23** (Part 2), pp. 1061–1074.
7. Lupan, O., Shishiyanu, S., Chow, L., Shishiyanu, T. (2008). Nanostructured zinc oxide gas sensors by successive ionic layer adsorption and reaction method and rapid photothermal processing, *Thin Solid Films*, **516** (10), pp. 3338–3345.
8. Vayssieres, L., Beermann, N., Lindquist, S.-E., Hagfeldt, A. (2001). Controlled aqueous chemical growth of oriented three-dimensional crystalline nanorod arrays: Application to iron(III) oxides, *Chem. Mater.*, **13** (2), pp. 233–235.
9. Postels, B., Kreye, M., Wehmann, H.-H., Bakin, A., Boukos, N., Travlos, A., Waag, A. (2007). Selective growth of ZnO nanorods in aqueous solution, *Superlattice Microst.*, **42** (1–6), pp. 425–430.
10. Pacholski, C., Kornowski, A., Weller, H. (2002). Self-assembly of ZnO: From nanodots to nanorods, *Angew. Chem. Intl. Ed.*, **41** (7), pp. 1188–1191.
11. Tian, Z.R., Voigt, J.A., Liu, J., Mckenzie, B., Mcdermott, M.J., Rodriguez, M.A., Konishi, H., Xu, H. (2003). Complex and oriented ZnO nanostructures, *Nat. Mater.*, **2** (12), pp. 821–826.
12. Vayssieres, L. (2003). Growth of arrayed nanorods and nanowires of ZnO from aqueous solutions, *Adv. Mater.*, **15** (5), pp. 464–466.
13. Lin, C.-C., Chen, H.-P., Chen, S.-Y. (2005). Synthesis and optoelectronic properties of arrayed p-type ZnO nanorods grown on ZnO film/Si wafer in aqueous solutions, *Chem. Phys. Lett.*, **404** (1–3), pp. 30–34.
14. Postels, B., Wehmann, H.-H., Bakin, A., Kreye, M., Fuhrmann, D., Blaesing, J., Hangleiter, A., Krost, A., Waag, A. (2007). Controlled low-temperature fabrication of ZnO nanopillars with a wet-chemical approach, *Nanotechnology*, **18** (19), art. no. 195602.
15. Postels, B. (2009). Niedertemperatur-Wachstum von ZnO-Nanosäulen für optoelektronische Anwendungen, Cuvillier, E; Auflage: 1, Göttingen.

16. Willander, M., Nur, O., Zhao, Q.X., Yang, L.L., Lorenz, M., Cao, B.Q., Zuiga Perez, J., Czekalla, C., Zimmermann, G., Grundmann, M., Bakin, A., Behrends, A., Al-Suleiman, M., El-Shaer, A., Che Mofor, A., Postels, B., Waag, A., Boukos, N., Travlos, A., Kwack, H.S., Guinard, J., Le Si Dang, D. (2009). Zinc oxide nanorod based photonic devices: recent progress in growth, light-emitting diodes and lasers, *Nanotechnology*, **20** (33), art. no. 332001.

17. Bakin, A., Behrends, A., Waag, A., Lugauer, H.-J., Laubsch, A., Streubel, K. (2010). ZnO-GaN hybrid heterostructures as potential cost-efficient LED technology, *Proc IEEE*, **98** (7), art. no. 5430882, pp. 1281–1287.

18. Wagner, A., Behrends, A., Waag, A., Bakin, A. (2011). Two step deposition method with a high growth rate for ZnO nanowire arrays and its application in photovoltaics, *Thin Solid Films* doi:10.1016/j.tsf.2011.10.206.

19. Postels, B., Bakin, A., Wehmann, H.-H., Suleiman, M., Weimann, T., Hinze, P., Waag, A. (2008). Electrodeposition of ZnO nanorods for device application, *Appl. Phys. A-Mater.*, **91** (4), pp. 595–599.

20. Bakin, A., Wagner, A., Schlenker, E., Postels, B., Al-Suleiman, M., Behrends, A., Elshaer, A.-H., Petukhov, V., Waag, A. (2008). ZnO nanostructures and thin layers for device applications, *ECS Trans.*, **16** (12), pp. 107–115.

21. Jimenez-Gonzailez, A.E., Nair, P.K. (1995). Photosensitive ZnO thin films prepared by the chemical deposition method SILAR, *Semicond. Sci. Tech.*, **10** (9), pp. 1277–1281.

22. Bakin, A., Che Mofor, A., El Shaer, A., Waag, A. (2007). Vapour phase transport growth of ZnO layers and nanostructures, *Superlattice Microst.* **42** (1–6), pp. 33–39.

23. Mofor, A.C., Bakin, A.S., Elshaer, A., Fuhrmann, D., Bertram, F., Hangleiter, A., Christen, J., Waag, A. (2007). Vapour transport growth of ZnO nanorods, *Appl. Phys. A-Mater.*, **88** (1), pp. 17–20.

24. Wagner, R.S., Ellis, W.C. (1964). Vapor-liquid-solid mechanism of single crystal growth, *Appl. Phys. Lett.*, **4** (5), pp. 89–90.

25. Huang, M.H., Wu, Y., Feick, H., Tran, N., Weber, E., Yang, P. (2001). Catalytic growth of zinc oxide nanowires by vapor transport, *Adv. Mater.*, **13** (2), pp. 113–116.

26. Li, S.Y., Lee, C.Y., Tseng, T.Y. (2003). Copper-catalyzed ZnO nanowires on silicon (1 0 0) grown by vapor-liquid-solid process, *J. Cryst. Growth*, **247** (3–4), pp. 357–362.

27. Zhang, Z., Wang, S.J., Yu, T., Wu, T. (2007). Controlling the growth mechanism of ZnO nanowires by selecting catalysts, *J. Phys. Chem. C*, **111** (47), pp. 17500–17505.

28. Rosina, M., Ferret, P., Jouneau, P.-H., Robin, I.-C., Levy, F., Feuillet, G., Lafossas, M. (2009). Morphology and growth mechanism of aligned ZnO nanorods grown by catalyst-free MOCVD, *Microelectr. J.*, **40** (2), pp. 242–245.

29. Perillat-Merceroz, G., Jouneau, H., Feuillet, G., Thierry, R., Rosina, M., Ferret, P. (2010). MOCVD growth mechanisms of ZnO nanorods, *J. Phys.: Conference Series*, **209**, art. no. 012034.

30. Zhang, B.P., Binh, N.T., Segawa, Y., Wakatsuki, K., Usami, N. (2003). Optical properties of ZnO rods formed by metalorganic chemical vapor deposition, *Appl. Phys. Lett.*, **83** (8), pp. 1635–1637.

31. Baxter, J.B., Aydil, E.S. (2009). Metallorganic chemical vapor deposition of ZnO nanowires from zinc acetylacetonate and oxygen, *J. Electrochem. Soc.*, **156** (1), pp. H52–H58.

32. Choopun, S., Vispute, R.D., Noch, W., Balsamo, A., Sharma, R.P., Venkatesan, T., Iliadis, A., Look, D.C. (1999). Oxygen pressure-tuned epitaxy and optoelectronic properties of laser-deposited ZnO films on sapphire, *Appl. Phys. Lett.*, **75** (25), pp. 3947–3949.

33. Sandana, V.E., Rogers, D.J., Hosseini Teherani, F., McClintock, R., Bayram, C., Razeghi, M., Drouhin, H.-J., Clochard, M.C., Sallet, V., Garry, G., Falyouni, F. (2009). Comparison of ZnO nanostructures grown using pulsed laser deposition, metal organic chemical vapor deposition, and physical vapor transport, *J. Vac. Sci. Technol B*, **27** (3), pp. 1678–1683.

34. Lorenz, M., Kaidashev, E.M., Rahm, A., Nobis, Th., Lenzner, J., Wagner, G., Spemann, D., Hochmuth, H., Grundmann, M. (2005). $Mg_xZn_{1-x}O$ ($0 \leq x < 0.2$) nanowire arrays on sapphire grown by high-pressure pulsed-laser deposition, *Appl. Phys. Lett.*, **86** (14), art. no. 143113, pp. 1–3.

35. Grundmann, M., Rahm, A., Nobis, T., Lorenz, M., Czekalla, C., Kaidashev, E.M., Lenzner, J., Boukos, N., Travlos, A. (2008). Growth and characterization of ZnO nano- and microstructures, in *Handbook of Self Assembled Semiconductor Nanostructures for Novel Devices in Photonics and Electronics*, 1st ed. (Amsterdam: Elsevier), Chapter 9, pp. 293–323.

36. Wagner, A., Bakin, A., Otto, T., Zimmermann, M., Jahn, B., Waag, A. (2011). Fabrication and characterization of nanoporous ZnO layers for sensing applications, *Thin Solid Films*, doi:10.1016/j.tsf.2011.10.210.

37. Lindström, H., Holmberg, A., Magnusson, E., Malmqvist, L., Hagfeldt, A. (2001). A new method to make dye-sensitized nanocrystalline solar cells at room temperature, *J. Photoch. Photobio. A*, **145** (1–2), pp. 107–112.

38. Lindström, H., Magnusson, E., Holmberg, A., Södergren, S., Lindquist, S.-E., Hagfeldt, A. (2002). A new method for manufacturing nanostructured electrodes on glass substrates, *Sol. Energ. Mat. Sol. C.*, **73** (1), pp. 91–101.

39. Bakin, A., El-Shaer, A., Mofor, A.C., Al-Suleiman, M., Schlenker, E., Waag, A. (2007). ZnMgO-ZnO quantum wells embedded in ZnO nanopillars: towards realisation of nano-LEDs, *Phys. Stat. Solidi* (*C*), **4**, n1, pp. 158–161.

40. Piechal, B., Yoo, J., Elshaer, A.-H., Mofor, A.-C., Yi, G.-C., Bakin, A., Waag, A., Donatini, F., Dang, L.S. (2007). Cathodoluminescence of single ZnO nanorod heterostructures, *Phys. Stat. Sol.* (*B*), **244** (5), pp. 1458–1461.

41. Heitsch, S., Zimmermann, G., Lenzner, J., Hochmuth, H., Benndorf, G., Lorenz, M., Grundmann, M. (2007). Photoluminescence of $Mg_xZn_{1-x}O$/ZnO quantum wells grown by pulsed laser deposition, *AIP Conference Proc.*, **893**, pp. 409–410.

42. Czekalla, C., Guinard, J., Hanisch, C., Cao, B.Q., Kaidashev, E.M., Boukos, N., Travlos, A., Renard, J., Gayral, B., Le Si Dang, D., Lorenz, M., Grundmann, M. (2008). Spatial fluctuations of optical emission from single ZnO/MgZnO nanowire quantum wells, *Nanotechnology*, **19** (11), art. no. 115202.

43. Cao, B.Q., Lorenz, M., Rahm, A., Von Wenckstern, H., Czekalla, C., Lenzner, J., Benndorf, G., Grundmann, M. (2007). Phosphorus acceptor doped ZnO nanowires prepared by pulsed-laser deposition, *Nanotechnology*, **18** (45), art. no. 455707.

44. Lorenz, M., Cao, B., Zimmermann, G., Biehne, G., Czekalla, C., Frenzel, H., Brandt, M., Von Wenckstern, H., Grundmann, M. (2009). Stable p-type ZnO:P nanowire/n-type ZnO:Ga film junctions, reproducibly grown by two-step pulsed laser deposition, *J. Vac. Sci. Technol.*, **27** (3), pp. 1693–1697.

45. Ohtomo, A., Tsukazaki, A. (2005). Pulsed laser deposition of thin films and superlattices based on ZnO, *Semicond. Sci. Technol.*, **20** (4), pp. S1–S12.

46. Lorenz, M. (2007). Pulsed laser deposition of ZnO-based thin films, in *Transparent Conductive Zinc Oxide: Basics and Applications in Thin Film Solar Cells* (Ellmer, K., Klein, A., Rech, B., ed.) Berlin: Springer, Vol. 104, pp. 305–360.

47. Park, W.I., Yi, G.-C. (2003). Fabrication and photoluminescent properties of ZnO/ZnMgO quantum structure nanorods, *Nanotechnology*, IEEE-NANO, 2003, **2**, 368–370.

48. Park, W.I., Yi, G.-C. (2004). Electroluminescence in n-ZnO nanorod arrays vertically grown on p-GaN, *Adv. Mater.*, **16** (1), pp. 87–90.

49. Zhang, X.-M., Lu, M.-Y., Zhang, Y., Chen, L.-J., Wang, Z.L. (2009). Fabrication of a high-brightness blue-light-emitting diode using a ZnO-Nanowire array grown on p-GaN thin film, *Adv. Mater.*, **21** (27), pp. 2767–2770.

50. Lee, J.Y., Lee, J.H., Seung Kim, H., Lee, C.-H., Ahn, H.-S., Cho, H.K., Kim, Y.Y., Kong, B., Lee, H.S. (2009). A study on the origin of emission of the annealed n-ZnO/p-GaN heterostructure LED, *Thin Solid Films*, **517** (17), pp. 5157–5160.

51. Titkov, I.E., Zubrilov, A.S., Delimova, L.A., Mashovets, D.V., Liniĭchuk, I.A., Grekhov, I.V. (2007). White electroluminescence from ZnO/GaN structures, *Semiconductors*, **41** (5), pp. 564–569.

52. Alivov, Y.I., Van Nostrand, J.E., Look, D.C., Chukichev, M.V., Ataev, B.M. (2003). Observation of 430 nm electroluminescence from ZnO/GaN heterojunction light-emitting diodes, *Appl. Phys. Lett.*, **83** (14), pp. 2943–2945.

53. Behrends, A., Bakin, A., Waag, A., Kwack, H.-S., Dang, L.S. (2010). Electroluminescence from a n-ZnO/p-GaN hybrid LED, *Phys. Status Solidi C*, **7** (6), pp. 1709–1711.

54. Sadaf, J.R., Israr, M.Q., Kishwar, S., Nur, O., Willander, M. (2010). White electroluminescence using ZnO nanotubes/GaN heterostructure light-emitting diode, *Nanoscale Res. Lett.*, **5** (6), pp. 957–960.

55. Alvi, N.H., Usman Ali, S.M., Hussain, S., Nur, O., Willander, M. (2011). Fabrication and comparative optical characterization of n-ZnO nanostructures (nanowalls, nanorods, nanoflowers and nanotubes)/p-GaN white-light-emitting diodes, *Scripta Mater.*, **64** (8), pp. 697–700.

56. Alvi, N.H., Riaz, M., Tzamalis, G., Nur, O., Willander, M. (2010). Fabrication and characterization of high-brightness light emitting diodes based on n-ZnO nanorods grown by a low-temperature chemical method on p-4H-SiC and p-GaN, *Semicond. Sci. Technol.*, **25** (6), art. no. 065004.

57. Kong, B.H., Cho, H.K., Kim, M.Y., Choi, R.J., Kim, B.K. (2011). InGaN/GaN blue light emitting diodes using Al-doped ZnO grown by atomic layer deposition as a current spreading layer, *J. Cryst. Growth*, **326** (1), pp. 147–151.

58. Fikry, M., Madel, M., Tischer, I., Thonke, K., Scholz, F. (2011). Luminescence properties of epitaxially grown GaN and InGaN layers around ZnO nanopillars, *Phys. Status Solidi A*, **208** (7), pp. 1582–1585.

59. Wang, S.-J., Li, N., Park, E.-H., Feng, Z.C., Valencia, A., Nause, J., Kane, M., Summers, C., Ferguson, I. (2008). MOCVD growth of GaN-based materials on ZnO substrates, *Phys. Status Solidi C*, **5** (6), pp. 1736–1739.

60. Kim, S.-Y., Lee, H.-J., Park, S.-H., Lee, W., Jung, M.-N., Fujii, K., Goto, T., Sekiguchi, T., Chang, J., Yao, T. (2010). Fabrication of a freestanding GaN layer by direct growth on a ZnO template using hydride vapor phase epitaxy, *J. Cryst. Growth*, **312** (14), pp. 2150–2153.

61. Drapak, I.T. (1968). Alloyed ZnO–Cu_2O heterojunction. In: *Semiconductors*, Vol. 2, p. 624.

62. Bakin, A., et al. (2011). *Oxide Semiconductors for Photovoltaic Applications*. 15th International Conference on II-VI Compounds; Mayan Riviera, Mexico, August 21–26 (invited talk).

63. Guerguerian, G., Elhordoy, F., Pereyra, C.J., Marotti, R.E., Martín, F., Leinen, D., Ramos-Barrado, J.R., Dalchiele, E.A. (2011). ZnO nanorod/CdS nanocrystal core/shell-type heterostructures for solar cell applications, *Nanotechnology*, **22** (50), art. no. 505401.

64. Kar, J.P., Jeong, M.C., Lee, W.K., Myoung, J.M. (2008). Fabrication and characterization of vertically aligned ZnMgO/ZnO nanowire arrays. *Mater. Sci. Eng. B*, **147**, pp. 74–78.

65. Wagner, A., Bakin, A., Waag, A. (2011). *Zinc Oxide based Dye Sensitized Solar Cells*. NTH School for Nanocontacts in Nanosystems, February 13th–19th, 2011, Bad Hofgastein, Austria.

66. Schlenker, E. (2009). *ZnO Based Nanostructures—Properties and Device Applicability*. Diss., TU Braunschweig, Cuvillier Verlag Göttingen, ISBN-10: 3867279675, ISBN-13: 9783867279673.

67. Mofor, A.C. (2007). *Fabrication and Characterisation of Device Quality ZnO Nanostructures*. Diss., TU Braunschweig, Cuvillier Verlag Göttingen, ISBN-10: 3867272867, ISBN-13: 9783867272865.

68. Elshaer, A.A.M. (2008). *Molecular Beam Epitaxy Growth and Characterization of ZnO-Based Layers and Heterostructures*. Diss., TU Braunschweig, Cuvillier Verlag Göttingen, ISBN-10: 386727701X ISBN-13: 9783867277013.

69. Kang, S.W., Mohanta, S.K., Kim, Y.Y., Cho, H.K. (2008). Realization of vertically well-aligned ZnO:Ga nanorods by magnetron sputtering

and their field emission behavior, *Crystal Growth Des.*, **8** (5), pp. 1458–1460.

70. SeongHo Jeong, Eray S., Aydil (2009). Heteroepitaxial growth of Cu_2O thin film on ZnO by metal organic chemical vapor deposition. *J. Crystal Growth*, **311**, pp. 4188–4192.
71. Akimoto, K., Ishizuka, S., Yanagita, M., Nawa, Y., Goutam K., Paul, T. S. (2006). Thin film deposition of Cu_2O and application for solar cells. *Solar Energy*, **80**, pp. 715–722.
72. Wang, R.-C., Hsin-Ying Lin (2010). Simple fabrication and improved photoresponse of ZnO–Cu_2O core–shell heterojunction nanorod arrays. *Sens. Actuators B*, **149**, pp. 94–97.

Chapter 2

Optical Properties of and Optical Devices from ZnO-Based Nanostructures

Michael Lorenz, Martin Lange, Christian Kranert, Christof P. Dietrich, and Marius Grundmann

Semiconductor Physics Group, Institut für Experimentelle Physik II, Universität Leipzig, Linnéstr. 5, D-04103 Leipzig, Germany

mlorenz@physik.uni-leipzig.de

The current state of optical properties of ZnO-based nanostructures is discussed in terms of optical recombination, of bandgap engineering and of nanoheterostructures such as core–shell quantum wells and quantum dots. Whispering gallery and Fabry–Perot modes are reviewed in the section on cavity effects, together with regular and random lasing in ZnO nanostructures. First results on Raman and infrared spectroscopy of nanostructures are listed. Promising optical device applications of ZnO nanowires include electrically pumped lasing arrays, photodetectors, and second harmonic generation.

2.1 Introduction

ZnO develops one of the richest families of different nanostructures, concerning first the huge variety of different structures and shapes,

Zinc Oxide Nanostructures: Advances and Applications
Edited by Magnus Willander

ISBN 978-981-4411-33-2 (Hardcover), 978-981-4411-34-9 (eBook)
www.panstanford.com

and second the corresponding piezoelectric, optical, and electrical properties [1–4]. The excellent luminescence properties of single selected ZnO nanowires were demonstrated by a full width of half maximum (FWHM) of the excitonic luminescence around 1 meV at 9 K, being close to best FWHM values of epitaxial thin films and near standard values of bulk single crystals [5]. First indications for stimulated emission at room temperature, for the basic concepts see [6], were provided for optically pumped ZnO nanorods by Yang, as mentioned in [1]. Quantum confinement effects in nanometer-thin ZnO nanobelts are obvious by a blue-shift of the photoluminescence peak [1, 7]. Whispering gallery modes in ZnO micro- and nanorods were demonstrated for the visible green emission band by Nobis [8, 9]. Later, whispering gallery mode lasing in the excitonic UV band was shown by Czekalla [10, 11] and others. First device applications of ZnO nanowires are reported only recently, for example photocurrent gas sensors and light-emitting diodes (LEDs) [2, 12]. Another promising application could be the near-surface light wave manipulation by engineered ZnO nanowire arrays [13]. Hierarchical structures made of networks of nanotubes and nanowires [14] or ZnO nanoplate-nanowire architectures [15] may open new directions for integrated device applications.

This chapter on optical properties of and optical devices from ZnO nanostructures is organized as follows: First, the optical recombination, bandgap engineering, and nanoheterostructures are discussed, and then cavity effects, including whispering gallery and Fabry–Perot modes, and regular and random lasing are reviewed. The state of Raman and infrared spectroscopy is shown, and finally the demonstration of first optical devices based on ZnO nanowires is reviewed. The latter include electrically pumped light-emitting diodes and lasers, photodetectors, and second harmonic generation devices.

2.2 Optical Recombination

The optical properties of ZnO-based nanostructures that can be grown by various growth techniques are strongly connected with other properties of the nanostructure itself. In this regard, luminescence measurements are often carried out, in order to conclude from the optical properties to other properties, e.g., crystal quality, defects, and dopants and their concentration. In addition, also optical properties like bandgap or index of refraction are of large interest. In the majority of the studies, measurements are performed at room temperature. Low-temperature or variable-temperature measurements are also reported from which more conclusions can be drawn. By applying these techniques, characteristic effects that change the nanostructures' optical properties, e.g., surface exciton recombination or quantum confinement in ultra-thin nanowires, can be studied.

It is essential to be able to tailor optical properties in order to increase the functionality of devices and their efficiency. Bandgap engineering is in this regard an important step to realize heterojunctions that can be applied in quantum wells and quantum dots. In this context the growth and optical properties of nanowire heterostructures, but for comparison also of thin films, made of ZnO and its ternary alloys with MgO, BeO, and CdO will be discussed.

2.2.1 Optical Recombination of ZnO Nanostructures

The optical recombination (spontaneous emission) typically consists of an UV-band near the ZnO bandgap energy and additional bands in the visible spectral range due to midgap defects and impurities. Without consideration of size and surface effects, the luminescence of ZnO nanostructures follows in general that of single crystalline ZnO bulk material or epitaxial films [5].

The low-temperature UV-emission of ZnO typically exhibits a number of recombination channels in the near-band edge (NBE) region as summarized in [16]. Several prominent recombination lines are observed in luminescence measurements of a single crystal performed at 12 K (see Fig. 2.1a). Besides the free exciton emission of the longitudinal and transversal exciton (indicated by A_L and A_T), donor-bound exciton emission is observed. The most intense lines can be attributed to A-excitons bound to neutral shallow donors (D^0X_A). In addition, excited states such as rotational and vibrational states, B-excitons bound to the donors (D^0X_B) or excited states occur [17]. Another recombination channel is given by A-excitons bound to ionized shallow donors (D^+X_A). The spectral range of the various emissions of ZnO is compiled in Fig. 2.1b. Further below the bandgap the predicted emission range of acceptor bound excitons (A^0X) [18] and deeply bound excitons (Y) [19] follows. Two electron satellites

(TES) of the D^0X_A and Y transitions are observed at even lower energies. In nano-size ZnO particles also surface exciton (SX)-related recombination channels are found. Only for some of the donors the chemical identity is known [18]. In addition to all the other UV-emission channels discussed before, also donor–acceptor pair recombination (DAP) can be observed [20].

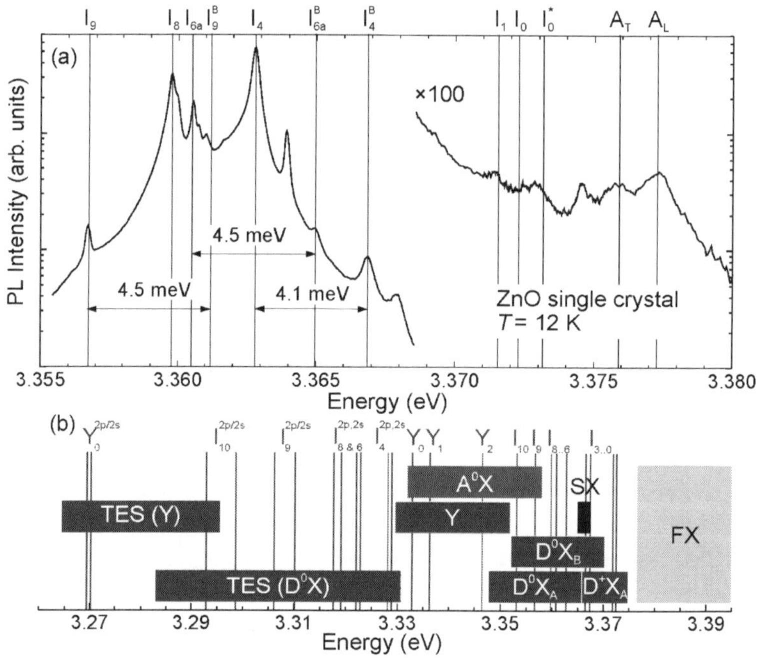

Figure 2.1 (a) Near band-edge luminescence of a ZnO single crystal with prominent recombination channels of ionized and neutral donor-bound exciton as well as free exciton emission at low temperatures (12 K). For the nomenclature of the transitions see [18]. (b) Spectral range of the emission of shallow-donor bound excitons (with TES), deeply bound excitons (with TES), acceptor bound excitons, surface excitons and free excitons. Adapted from [19].

All these processes exhibit longitudinal optical (LO) phonon replicas that are shifted by multiples of the LO-phonon energy of ~72 meV to lower energies. The distribution on the zero-phonon transition and the replicas is determined by the Huang-Rhys factor *S* [21, 22] and is Poisson-like. Therefore, the intensity of the replicas depends on the coupling strength of the electronic transition to the longitudinal optical polarization field. As a consequence of different coupling strengths the intensity and therefore also the number of observed LO-phonon replicas can vary, for example, for (D^0X_A) and Y recombination channels [19].

The low-temperature UV luminescence of ZnO nanostructures is despite some special phenomena in general similar to that of bulk material. Therefore, bound and free excitons and the features as discussed before are typically observed. Low-temperature PL spectra can be found, e.g., in [23–30]. Figure 2.2 shows PL spectra of ZnO NWs measured at 11 K, exhibiting emission of neutral-donor bound, surface and free excitons. The most intense observed emission originates from the I_9 and I_6 lines that can be ascribed to In or Al donors, respectively, which are expected to originate from the substrates [23]. In general, the number of observed recombination lines of donor-bound excitons is smaller for NWs in comparison to single crystals as result of larger linewidth (in some NW samples) but also due to a smaller number of different impurities. The enhanced linewidth that is observed in some studies can be explained by varying donor concentrations within single NWs [31] or by a statistics effect of the NW-ensemble.

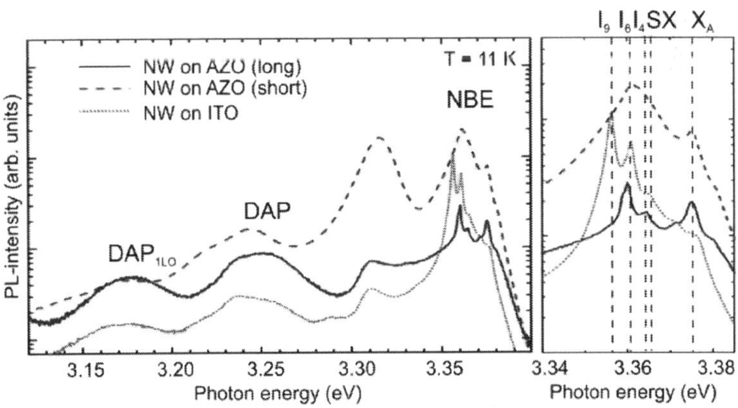

Figure 2.2 Low-temperature PL spectra of ZnO NWs grown on different substrates with donor-bound, surface and free exciton emission in the NBE region. AZO stands for Al-doped ZnO thin films, and ITO for indium tin oxide. Donor–acceptor pair recombination is observed at 3.235 eV with respective LO-phonon replicas. Reprinted from [23] with permission. Copyright 2011 American Chemical Society.

One important aspect in the NW UV luminescence is that in some samples the intensity of surface bound exciton emission (3.365 eV–3.368 eV) is strongly enhanced due the high surface-to-volume ratio, such that in some cases it is the strongest recombination channel in the UV-range, see Fig. 2.3 and in addition [33–38]. The intensity of the surface exciton (SX) peak relative to the donor-bound exciton emission strongly depends on several parameters: (i) the size of the NW (smaller objects show a more pronounced SX-peak [34]; (ii) the excitation conditions [35] such as operation mode, wavelength and intensity; and finally (iii) also the sample temperature [32–37].

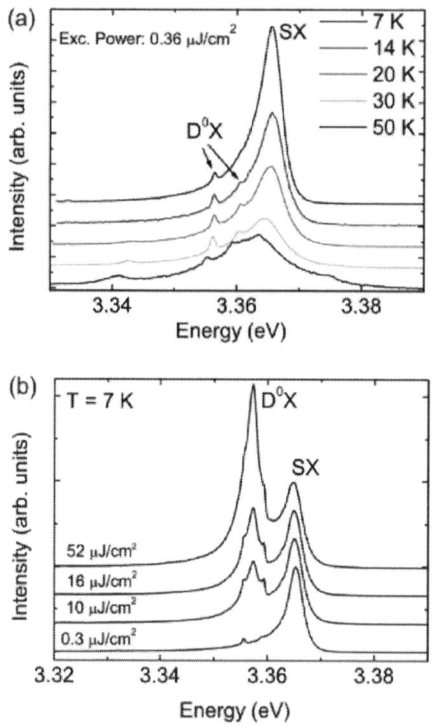

Figure 2.3 PL spectra of a nanowire/nanowall ensemble taken at different temperatures in (a) and at different excitation fluences in (b). Reprinted from [32] with permission. Copyright 2006 by The American Physical Society.

The microscopic origin of the SX peak is still under discussion. Biswas et al. could provide clear evidence that the responsible defects are adsorbed surface species [38]. Their final conclusion is that the SX peak occurs due to excitons bound at the ZnO surface that is modified by an adsorbate, most likely an OH-related species [38]. However, further research is necessary to understand the detailed microscopic origin. To obtain further insight into the origin of the SX-emission, time-resolved luminescence measurements were performed, see [32, 39]. The authors propose that their experimental observations result from weakly and strongly localized excitons in a surface-near layer with a thickness between 20 and 30 nm [39]. Additionally, one has to keep in mind that the intensity of the SX recombination band is strongly connected to the detailed nanostructure morphology and not only to the surface-to-volume ratio [38].

Another effect that can be observed as consequence of the small size of NWs with diameters below 10 nm, is the blue-shift of the excitonic emission due to quantum confinement. Therefore, these wires are often called "quantum wires."

There are two different approaches in order to calculate the blue-shift as function of the quantum wire diameter. It can be determined by applying an effective-mass (EMA) based envelope function theory, as done by Xia et al. who performed six-band calculations for NW diameters $d \geq 3$ nm [29]. An alternative approach is given by density functional theory (DFT) as performed by Li et al. for $1.5 \leq d \leq 4$ nm [40] and Tuoc for $0.9 \leq d \leq 3$ nm [41]. For the DFT calculations one has to keep in mind that the results are in general more qualitative than quantitative and that there are still problems even in calculating the bandgap energy for bulk material only. However, all calculations show the expected behavior of an increasing blue-shift with decreasing NW-diameter, but the quantitative results of the calculations differ significantly. The calculated values can be compared with experimental ones, as discussed in the following.

For experimental determination of the blue-shift of the bandgap energy, several emission and adsorption measurements were performed [34, 42–44]. For NW diameters between 4 and 12 nm large blue-shifts of up to 273 meV are reported (see Fig. 2.4). However, if plotted as function of the NW diameter, the difference between the data points is very different for some reported values. The ones by Park et al. [44] and Wang et al. [42] agree well with each other and with the predicted values from the theoretical calculations discussed above. The NWs studied by Stichtenoth et al. [34] and Greyson et al. [43] deviate from the observed trends of the other experimental values and the theoretical curves. The blue-shift is too *small* for the *thin* NWs [34] and too *large* for the *thicker* NWs [43], respectively. Finally, in the studied NW diameter range the EMA-based theory agrees best with the reasonable experimental values reported in [42, 44]. However, further and more detailed experimental and theoretical studies are necessary in order to improve the understanding of the blue-shifts that occur in quantum wires.

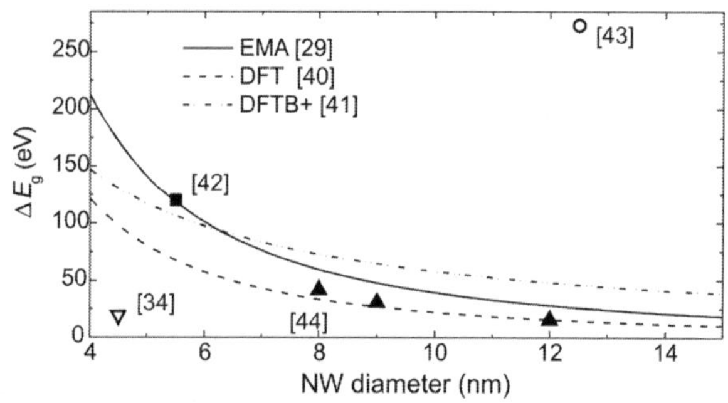

Figure 2.4 Increase of the bandgap energy vs. diameter of the NW. Experimental values are given as data points after Refs. [34, 42–44] and compared with theoretical curves using EMA [29] and DFT [40, 41] calculations. The curves are extrapolated beyond the calculated range (for calculation range see text).

With increasing temperature the luminescence peaks of the donor-bound and free excitons show a characteristic red-shift due to a reduction of the fundamental ZnO-bandgap, see for example [24, 26–30]. Additionally, the intensity of the donor-bound excitons and their phonon replicas quenches as result of the delocalization of excitons. In general, at a certain temperature the transition from dominating donor-bound to free exciton emission occurs. Its value depends on the specific sample properties, as both, chemical identity of the present donors and their concentration, directly influence the localization energy of the exciton to the donor and the capture cross sections of the donor. As the type of donors and their concentration varies for different nanostructures and growth techniques, there exists a wide range for the transition temperature.

At room temperature the recombination spectra are dominated by free excitons. An effect that is typical for nanostructures is that different NBE emission maxima of different types of nanostructures with peak energies between 3.12 and 3.30 eV are observed [45]. One of the reasons is given by the fact that the maximum of the room temperature emission band is a superposition of the zero-phonon transition and the first phonon replica of the free exciton and that in some cases the spectra are even dominated by the first phonon replica due to the bottleneck effect [28]. Another reason for the spectral shifts can be given by different defect concentrations in the various nanostructures with different size and surface-to-volume ratios [45]. Also the detection and excitation geometry can influence the observed luminescence, although the resulting peak shifts are relatively small [42].

Besides the NBE emission in the UV-spectral range the spectra often exhibit a number of different deep peaks in the visible range, which are attributed to defects. The most commonly observed band in ZnO is the green one, but yellow, orange, blue, and violet emissions are also observed. For an overview, see [45]. The defect-related emission can be observed at low temperatures but also up to room temperature and above. The origins of the emission bands with characteristic energies are in some cases still discussed controversially [45].

In general, it is difficult to directly conclude on the sample quality from the luminescence measurements, as the absolute intensity and the intensity ratios of the near band edge and defect emission strongly depends on the excitation conditions, i.e., intensity, wavelength, operation mode, and geometry. Therefore, the sample quality is not directly connected with the ratio of ultraviolet and defect emission. However, under comparable experimental conditions, relevant conclusions can be drawn. A better criterion for the sample quality can be given by the presence of the free exciton emission and the spectral distinction of A and B exciton [45, 46].

2.2.2 Bandgap Engineering

The rising functionality and efficiency of many modern solid-state devices is based on the appropriate usage of semiconductor heterostructures. In this regard, they are applied in light-emitting diodes, solid-state lasers, and high-mobility field-effect transistors. To tune the parameters of the heterojunction, the possibility to modify the properties of the corresponding materials in the desired way is indispensable. Bandgap engineering of the wurtzite ZnO gives the possibility to tune the bandgap and is mandatory for the realization of ZnO-based quantum wells and quantum dots. To adjust the fundamental bandgap, Zn atoms are in generally substituted by Mg or Be atoms to increase the bandgap, or with Cd atoms to reduce it. By substituting Zn atoms with Mg, Be, or Cd atoms also other properties, e.g., lattice constants or crystal structure, can change. Important material properties of the respective binary materials are summarized in Table 2.1.

Table 2.1 Material properties of ZnO and oxides that form ternary alloys with ZnO

	Crystal structure	Bandgap energy (eV)	*a*-Lattice constant (Å)	*c*-Lattice constant (Å)
ZnO	Wurtzite	3.37	3.2501[T1]	3.2071[T1]
MgO	Rocksalt	7.77[T2]	4.2123[T3]	—
BeO	Wurtzite	10.585[T4]	2.6979[T5]	4.3772[T5]
CdO	Rocksalt	2.42[T6]	4.6942[T7]	—

Note: T1, [47]; T2, [48]; T3, [49]; T4, [50]; T5, [51]; T6, [52]; T7, [53].

In Fig. 2.5 the bandgap energy is plotted over the a-lattice constant for ZnO, MgO, BeO, and CdO together with the bandgap energies of the respective solid solution thin films as function of their composition. For thin film ZnO-based ternary alloys, a wide spectral range from 1.85 to 5.37 eV is accessible in the wurtzite regime. The slope of the emission energy over composition is comparable for Mg and Be, but the maximum Be content that has been achieved is significantly larger, so that the largest emission energies are obtained for (Be,Zn)O.

Up to now, there exist a much smaller number of reports on nanostructures made of ZnO-based ternary alloys in comparison to thin films. Additionally, the accessible spectral range is significantly smaller. NWs with an enhanced bandgap energy were up to now exclusively reported for (Mg,Zn)O but not for (Be,Zn)O. Beside the fact that such NWs emit at higher energies they can additionally be utilized for the growth of nanowire (NW) heterostructures. A large number of approaches have been developed in order to obtain (Mg,Zn)O NW with large Mg contents and high crystal quality. The first reports were given by Heo et al. using molecular beam epitaxy (MBE), who reported a blue-shift of 60 meV [60]. Subsequently larger blue-shifts due to larger Mg contents were obtained by using metal organic vapor phase epitaxy (MOVPE) [61–63] or high-pressure pulsed laser deposition (hp-PLD) [64]. The largest Mg contents were reported for samples that were fabricated by using the vapor phase transport (VPT) method [65–68]. Emission energies as high as 4.07 eV—but still 0.4 eV below the thin film energies—are possible [65]. A drawback is that the VPT-grown samples only exhibit a quasi-alignment. Besides, (Mg,Zn)O NWs with a blue-shift of 150 meV were fabricated using a low-temperature hydrothermal approach [69] or a shift of 200 meV via in-diffusion of Mg into ZnO using ZnO cores and MgO shells [70].

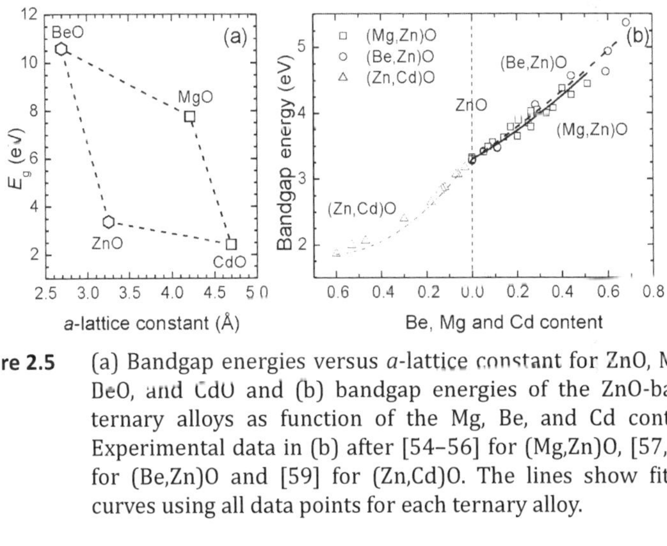

Figure 2.5 (a) Bandgap energies versus *a*-lattice constant for ZnO, MgO, BeO, and CdO and (b) bandgap energies of the ZnO-based ternary alloys as function of the Mg, Be, and Cd content. Experimental data in (b) after [54–56] for (Mg,Zn)O, [57, 58] for (Be,Zn)O and [59] for (Zn,Cd)O. The lines show fitting curves using all data points for each ternary alloy.

An emission energy smaller than that of ZnO was successfully obtained in (Zn,Cd)O NWs. Also for the solid solutions of ZnO and CdO the most common growth technique is the VPT method [67, 68, 71]. Chu et al. report that they were able to tune the emission energy of their (Zn,Cd)O NWs in a wide spectral range from 2.08 to 3.32 eV (for the case of ZnO) or even 4.00 eV (for the case of (Mg,Zn)O [68]. Much smaller Cd contents and red-shifts of up to 0.3 eV were obtained for samples of nanoparticle aggregates that were grown by an electrochemical approach [72]. By oxidizing (Zn,Se)O NWs, it is also possible to obtain (Zn,Cd)O NWs, but up to now the largest reported red-shift is as small as 0.05 eV [73].

The emission energies of (Mg,Zn)O and (Zn,Cd)O NWs for the various growth techniques are summarized in Fig 2.6. The fitting curves of the emission energies for (Mg,Zn)O and (Zn,Cd)O bulk material (see Fig. 2.5) are added for comparison. For (Mg,Zn)O a good agreement is obtained, although some data points significantly deviate from the fitting curve. However, for (Zn,Cd)O there is a large number of samples which exhibit significantly higher energies than expected from the Cd content, which might be caused by an imprecise determination of the Cd content.

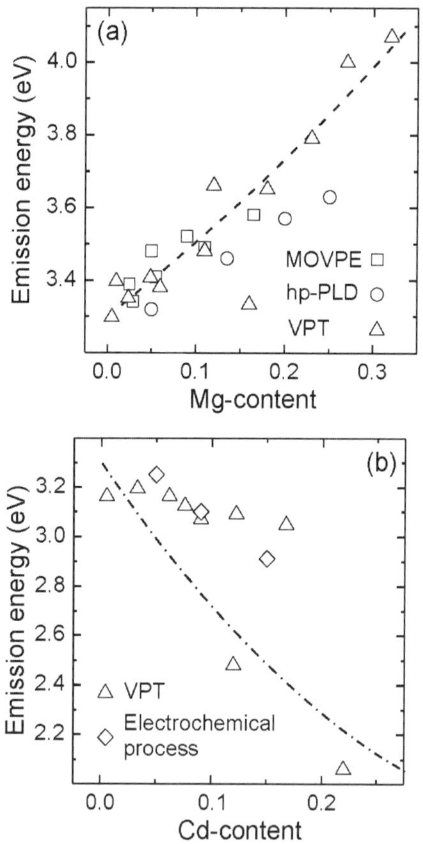

Figure 2.6 (a) and (b) Emission energy of (Zn,Mg)O and (Zn,Cd)O NWs versus Mg- and Cd content, respectively. Experimental data are after [61–68, 79, 74] for (Mg,Zn)O (a) and after [68, 71, 72, 75, 76] for (Zn,Cd)O (b). For comparison the lines that were fitted to experimental data of the thin films (see Fig. 2.5), are given.

To summarize, also in NWs the bandgap energy and therefore the emission energy can be tuned by substituting Zn atoms with Mg or Cd atoms in a wide range from 2.08 eV (yellow) to 4.08 eV (ultraviolet). However, the accessible spectral range is smaller for NWs in comparison to thin films as result of constrictions during NW growth. The (Mg,Zn)O can be the basis for the growth of NW heterostructures in combination with ZnO or (Zn,Cd)O quantum wells or quantum dots as active material.

2.2.3 Nanowire-Heterostructures: Quantum Wells and Quantum Dots

The successful growth of NW heterostructures is an important step in order to enhance the emission properties of NWs. The incorporation of quantum wells (QWs) and quantum dots (QDs) offers the possibility of a tunable emission energy and a better performance of optical devices in comparison with active bulk layers and are therefore an important step to realize photonic NW devices [12, 77, 78]. The efficiency of the device is strongly connected to challenges in the growth of the NW heterostructure. To have access to a wide spectral range large Mg- and Cd contents are essential. However, also exciton diffusion and carrier transport from the barrier layers to the QWs/QDs has to be efficient and therefore plays a key-role. Thus, reasonable substituent content is limited. Another important aspect is strain-management which also limits the maximum contents, as the lattice mismatch between the ternary ZnO itself and its ternary alloys, see Fig. 2.5, increases with increasing Mg, Be and Cd content.

The growth of core/shell NW heterostructures with a (Mg,Zn)O NW core has not been reported up to now. VPT processes, by which NWs with large Mg contents can be grown, are usually carried out in a tube furnace, such that the source material cannot be easily changed during growth. Therefore, the fabrication of heterostructures and QW structures is challenging since different process runs and thickness control down to nanometers are essential. Directed growth methods like MBE or PLD are up to now limited to minimal (Mg,Zn)O NW densities of 1–10 μm^{-2}, but densities below 0.1 μm^{-2} are necessary to avoid shadowing effects and obtain homogeneous core/shell heterostructures [79]. In this context, new basic approaches have to be developed for the growth of NW heterostructures before (Mg,Zn)O NW can be applied as core structure. Therefore an alternative approach is applied by using ZnO NW cores for the heterostructure growth. They show a high crystalline quality and can be grown with a low lateral density.

The most simple heterostructure is given by a ZnO core covered with a (Mg,Zn)O shell. Such structures can be realized, for example, by a two-step thermal evaporation process as reported by Kong et al. [80, 81]. A blue-shift of the (Mg,Zn)O-related luminescence of about 0.13 eV in comparison to the ZnO is observed [81]. Larger blue shifts up to about 0.2 eV can be obtained for samples grown by using PLD [79] or MOVPE [82]. In principle also larger Mg contents can be obtained and were applied as barrier layers, e.g., in quantum well nanowire heterostructures with blue-shifts up to 0.55 eV by using PLD [83]. However, in this study no ZnO/MgZnO core–shell single heterostructures were studied but solely quantum well heterostructures.

For optoelectronic devices the incorporation of QW/QDs is necessary. The modulation of the composition can be realized with a modulated composition along defined directions, either the radial and/or the axial direction [84]. QWs with emission energies above that of ZnO can be realized by surrounding ZnO NWs with a ZnO/(Mg,Zn)O QW shell.

Radial QW heterostructures of that type were grown using metal-organic vapor phase epitaxy (MOVPE) and optically investigated by Yi's group [82, 85]. Using MOVPE randomly oriented NWs were covered homogeneously with QWs, even in high density, because the reactant molecules can homogeneously cover the major part of the NW. The emission energy of the QWs was tuned between 3.382 and 3.467 eV.

The group of G. C. Yi also realized MOVPE-grown QWs on NWs with an axial orientation [7, 86–88], see Fig. 2.7. In this approach, well-aligned high density ZnO NWs served as core and the ZnO/(Mg,Zn)O QWs were grown in axial direction. As result of the parallel orientation and of the homogeneous length of the NWs the QWs were solely grown on the NWs' tips. By changing the QW thickness the emission energy was tuned between 3.375 and 3.515 eV. Bakin et al. realized axially oriented QWs on well-aligned NWs by applying MBE. The NWs were grown by a VPT-process on SiC/sapphire templates [12, 89]. The QW-related emission was observed at 3.42 eV.

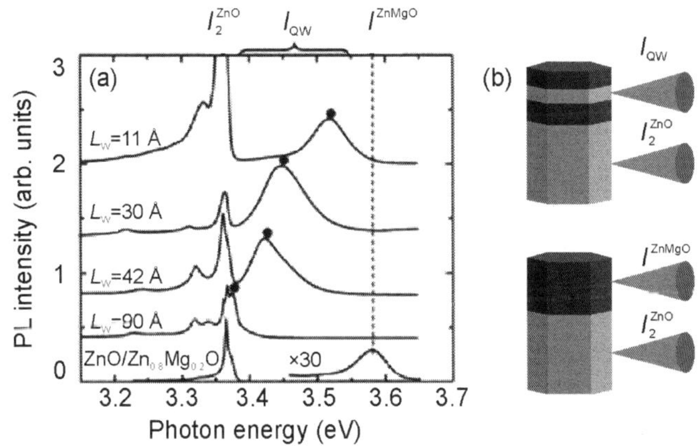

Figure 2.7 (a) PL spectra (10 K) of an axial ZnO/(Mg,Zn)O core–shell heterostructure and ZnO NW cores with ZnO/(Mg,Zn)O QW shells with different QW thicknesses. Reprinted from [7] with permission. Copyright 2004 American Chemical Society. (b) Schematic of the axial heterostructures with emission origins (after [7]).

Core–shell heterostructures with a complete ZnO/(Mg,Zn)O QW shell are inherently composed of a QW in axial and radial direction. Such structures can be grown, for example, by a two-step PLD process, as first reported by Cao et al. [79]. To obtain homogenous QWs, the NW cores were grown with a low lateral density below 0.1 μm^{-2}. If the growth rates in axial and radial direction were equal, the emission of the axial and radial QW would occur at the same energy. Cao et al. could prove that for a highly directed growth technique like PLD this is not the case. The axial growth rate is about a factor of 3–5 larger than that in radial direction, so that the emission of the axial and radial QWs is observed at different energies [79]. Depending on the QW thickness the emission occurred at energies of about 3.4 eV and between 3.42 and 3.47 eV for the axial QW and the radial QW, respectively.

However, all of the reported results obtained relatively small blue-shifts of the QW luminescence of up to 0.14 eV due to quantum confinement. This is result of the relatively small Mg contents in the barrier layer, although QW thicknesses down to 1.1 nm were applied, see Fig. 2.8. In this regard, higher Mg contents are essential in order to obtain higher QW emission energies. Following this approach, recently, NW core/shell heterostructures with large confinement energies were reported by Lange et al. [83]. Their core–shell samples were also grown by a two-step PLD process but barrier layers with Mg contents exceeding 25% were achieved. The QW emission energy was thus tuned in a much larger spectral range from 3.42 to 3.68 eV, see Fig. 2.8. For future device applications, a laterally homogeneous emission of the QWs is essential. In this regard, Lange et al. studied the homogeneity of the emission of the radial QW along the *c*-axis of the NW. Besides a significant increase near the tip, which was ascribed to a locally reduced growth rate due to particle diffusion during growth, a homogeneous energy of the QW luminescence with a standard deviation of only 4 meV was observed [83].

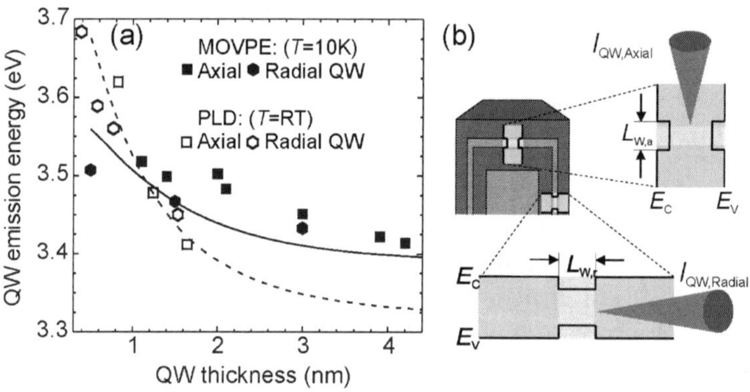

Figure 2.8 (a) QW emission energies of axial and radial QWs grown on ZnO NWs by using MOVPE and PLD. Experimental data for MOVPE after [7, 82] measured at 10 K and for PLD after [83] measured at room temperature (RT). The lines are theoretical calculations for both cases. The applied parameters can be found in [90]. (b) Schematic of a ZnO/(Mg,Zn)O QW core/shell heterostructure with band offsets and emission origins of the axial and radial QW.

QWs with emission energies below that of bulk ZnO can be realized by surrounding ZnO NWs with a (Zn,Cd)O/ZnO QW shell. The growth of such a structure is in principle less challenging as high quality ZnO-cores can be applied and as the NW has to be covered only by two layers. However, for most of the growth techniques it is relatively hard to obtain large Cd contents. Additionally, the ZnO nanostructures usually exhibit an intense green emission, which is in the same spectral range as that of the QW and which might be more intense than the QW luminescence. Experimental results were first reported by Cheng et al. [91]. For the growth of the (Zn,Cd)O coaxial MQW nanowire arrays they combined chemical vapor deposition (CVD) with PLD and applied (Zn,Cd)O with a Cd content of 4%. The luminescence of the MQWs is observed at 3.2 eV and is therefore 0.16 eV red-shifted relative to the ZnO emission energy, see Fig. 2.9. The corresponding (Zn,Cd)O bulk emission is observed at 3.11 eV, so that a quantum confinement is proven.

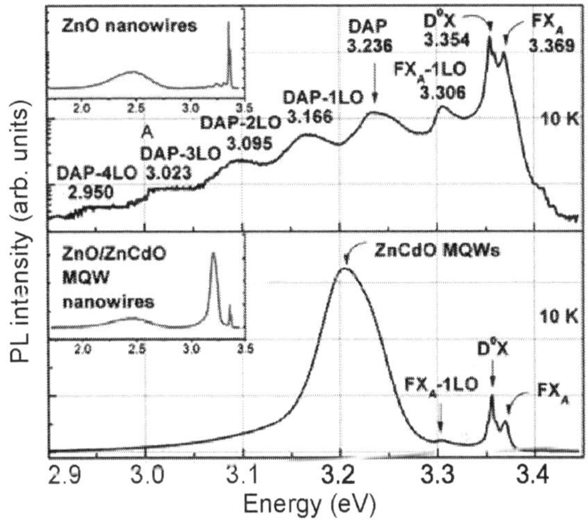

Figure 2.9 PL spectra of *top* ZnO NW arrays in comparison to *bottom* (Zn,Cd)O/ZnO MQW NW arrays, both measured at T = 10 K. Reprinted from [92] with permission. Copyright 2010 American Chemical Society.

An interesting approach is the incorporation of QDs into the NW structures. The optical and electrical properties of QDs allow the realization of, for instance, single-photon sources or quantum dot lasers. QD-like emission in ZnO-based NW heterostructures has been up to now only reported by Czekalla et al. [93]. The NW heterostructures were grown by an ex situ two-step PLD process, with the heterostructure in axial direction of the NW axis using a growth sequence comparable to that for the QWs. Spatially resolved luminescence measurements at low temperatures revealed that the emission band of the heterostructure was only observed at the top of the NW heterostructure and was composed of several sharp emission lines [93]. By changing the position of the excitation beam the sharp lines sensitively fluctuate. From these results, together with TEM results, Czekalla et al. concluded that quantum-dot-like traps for excitons are formed in laterally strained areas and ZnO fluctuations in the (Mg,Zn)O barrier layers on the NW [93].

In summary, the successful incorporation of ZnO-based QWs and QDs in ZnO NW heterostructures is possible, as discussed above. Emission energies above that of ZnO were obtained in axial-, radial- or complete core–shell QW heterostructures with (Mg,Zn)O barriers. In comparable heterostructures also, a QD-like emission was observed in low-temperature luminescence measurements. By applying (Zn,Cd)O/ZnO QWs, emission energies below that of ZnO were realized. However, the energetic range in which the ZnO-based QW/QD emission energy was tuned is relatively small (3.17–3.68 eV) in comparison to the accessible range for thin films (1.76–5.37 eV) and NWs (2.06–4.07 eV). This can be explained by the advanced challenges in the growth of NW heterostructures and of the alloys on the NWs. Additionally, there is further scope to improve the structures, as up to now topics such as lattice-matched structures have not been tackled.

2.3 Cavity Effects

ZnO nanostructures in air embody natural, perfectly built dielectric resonators. Due to the high index contrast, they are able to trap light effectively. The angle for total internal reflection between ZnO and air is below 30° for energies larger than 2 eV. This allows a variety of stable optical pathways inside ZnO nanostructures that can generally be classified into two types of cavity modes: whispering gallery modes (Section 2.3.1) and Fabry–Perot modes (Section 2.3.2). When the structure size exceeds a critical value and strong light scattering occurs, even randomly proceeding stable pathways can develop possibly leading to so-called "random lasing" (Section 2.3).

2.3.1 Whispering Gallery Modes

Optical whispering gallery modes (WGM) got into the focus of interest in the early nineties when the first microdisk resonator based on the InP/InGaAsP material system with circular cross section [94] was successfully fabricated and stimulated emission of cavity modes propagating circumferentially was demonstrated.

In 1912, WGM were initially mentioned by Lord Rayleigh who tried to explain the origin of extraordinary acoustic effects in St. Paul's cathedral in London [95]. In general, WGM arise due to constructive interference of light waves that propagate in a cross section plane with C_m symmetry ($m \geq 2$) by total internal reflection at the plane boundaries. The appearance of distinct WGM signatures is strongly correlated to the sample morphology and crystal structure they propagate in.

All ZnO nanostructures in which WGM have been observed so far exhibit cross sections that reflect the hexagonal wurtzite crystal structure of ZnO. Up to now, WGM could be observed in ZnO structures with triangular (m = 3) [97], hexagonal (m = 6) [8] or dodecagonal (m = 12) [98] cross section whereby the hexagonal structure is the most common one as measured by the number of publications. Note that even microtubes are able to exhibit WGM signatures [99]. Figure 2.10 shows possible closed pathways of WGM (red lines) that can generally occur in a nanostructure with regular triangular (A–B), hexagonal (C–E), or dodecagonal cross section (F–J) [96], respectively. The different types of WGM differ by the length of their closed optical pathway L, the angle of incidence θ_i of impinging wave fronts on the cross section boundary and the number of reflections M at the boundary that give rise to phase shifts of the light waves for each total reflection.

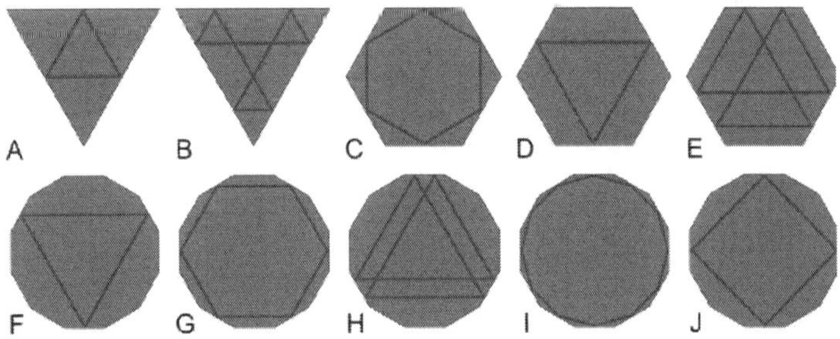

Figure 2.10 Schematic draw of possible closed pathways (red lines) in regular triangles, hexagons, and dodecagons, respectively.

All modes can unambiguously be identified by their mode energies for a given cavity cross section width d (in our case the longest possible distance between two cross section corner points) and the ZnO refractive index n. Approximately, these mode energies E can be calculated via a simple, geometrical plane wave picture that accounts for constructive interference of light waves that pass the length of a closed roundtrip L and occurring phase shifts at the boundary due to total internal reflection:

$$E = \frac{hc}{nL}\left\{N + \frac{M}{\pi}\arctan\left[\kappa\sqrt{\frac{n^2\sin^2\theta_i - 1}{\cos^2\theta_i}}\right]\right\} \tag{2.1}$$

with Planck constant h, speed of light c and mode number N. Further, κ accounts for the birefringence of ZnO and is $\kappa = n$ for TE- and $\kappa = 1/n$ for TM-polarization. Values for L (in units of d), M, and θ_i can be found in Table 2.2 for different types of WGM and resonator cross sections from Figs. 2.10A–J. We have to annotate that WGM with $\theta_i = 30°$ are only stable for refractive indexes $n \geq 2$ which is roughly fulfilled for energies above 2 eV.

Table 2.2 Length L (in units of cross section width d), number of total reflections M and angle of incidence θ_i for different types of WGM and resonator cross sections (letters A–J according to Fig. 2.10)

	Triangular		Hexagonal			Dodecagonal				
	A	B*	C	D	E*	F	G	H*	I	J
L/d	$\frac{3}{2}$	3	$\frac{3\sqrt{3}}{2}$	$\frac{9}{4}$	$\frac{9}{2}$	$\frac{3(\sqrt{3}+3)}{4\sqrt{2}}$	$\frac{3(\sqrt{3}+1)}{2\sqrt{2}}$	$\frac{3(\sqrt{3}+1)}{2\sqrt{2}}$	3	$\sqrt{3}+1$
M	3	6	6	3	6	3	6	6	12	4
θ_i	30°	30°	60°	30°	30°	30°	60°	30°	75°	45°

*Note that modes B, E, and H in Fig. 2.10 already constructively interfere for half the optical pathway and half the number of reflections and thus exhibit different mode energies in Fig. 2.10, see Ref. [96].

However, as Eq. (2.1) is derived from a simple plane wave model, its applicability and validity is limited to cross section widths much larger than the incident wavelength ($d >> \lambda$) [100]. Since most common experiments are carried out with UV excitation, Eq. (2.1) can roughly be applied only for samples with $d > 2$ μm [11]. Experimentally, the lowest possible WGM with mode number $N = 1$ has been observed in a tapered nanowire with about 70 nm diameter, which is much smaller than the excitation wavelength λ [8], and later also by [101]. For this case, Eq. (2.1) can still be used for a qualitative description of the mode behavior, but for a quantitative understanding, a more detailed calculational approach is needed. Two suitable approaches for the theoretical treatment of WGM in nano-sized ZnO structures are (i) Wiersig's boundary element method (BEM) [9, 100] and (ii) finite difference time domain (FDTD) simulations established by Taflove and Brodwin [102] in the late seventies.

(i) Wiersig presented the BEM in Ref. [100], its experimental applicability was shown by Nobis et al. in Ref. [9] using a slightly refined algorithm. The BEM requires the discretization of the hexagon boundary in finite elements and solving the two-dimensional Helmholtz equation with wave function Ψ and its wave vector k

$$-\nabla\Psi^2 = n^2k^2\Psi \tag{2.2}$$

for every single discretized boundary element with appropriate, steady boundary conditions. The BEM procedure requires a certain flattening of the hexagon corner [9, 100] in order to avoid discontinuities and to save calculational time. A similar approach was presented by Pavlovic et al., who parameterized the hexagon cross section by approximating it by a linear interpolation between an ideal hexagon and a circle [103]. They solved Eq. (2.2) for non-vanishing wave vectors, i.e., equal to non-normal emission directions, by applying linear combinations of cylindrical harmonics to fulfill the boundary conditions for tangential electric and magnetic fields at the cross section boundary. This approach even allowed the calculation of photonic dispersion curves of single WGM.

(ii) Another way to calculate WGM for small nanostructures is to perform FDTD simulations [102]. They are based on space and time discretizations of the time-dependent Maxwell's equations. Electric and magnetic field vectors are solved iteratively in time and space until the transient is fully developed. Figure 2.11 shows normalized, TM-polarized electric field intensity distributions of low WGM mode numbers (N = 5–10) in a regular triangle, hexagon, dodecagon and circle, respectively, derived from FDTD simulations. The index of refraction was chosen to be n = 2.5, which is roughly equal to the ZnO index of refraction in the excitonic regime

(3.25 eV) [104]. Note that independently from the chosen method the solution should always give the same result as depicted in Fig. 2.11.

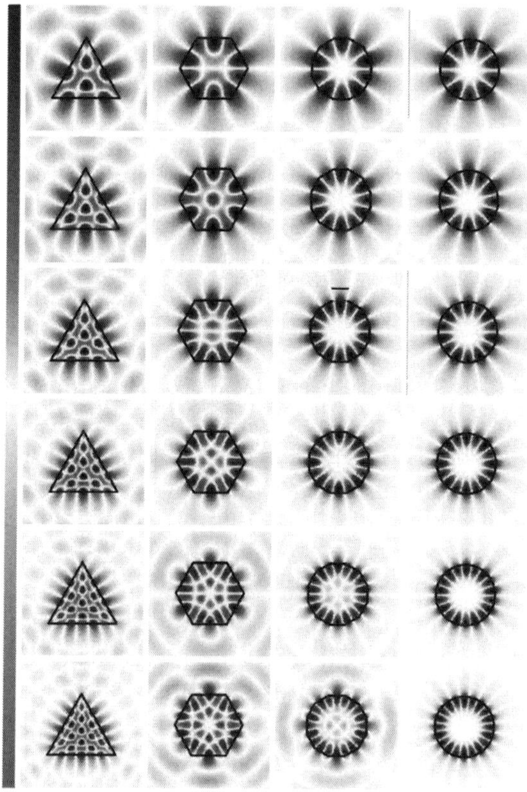

Figure 2.11 Normalized, TM-polarized electric field strength distributions of mode numbers 5 (top) –10 (bottom) in a regular triangle (left), hexagon (middle left), dodecagon (middle right) and circle (right) derived from FDTD simulations. The index of refraction was set to $n = 2.5$.

WGM can adopt very high quality factors, which are a prerequisite for the achievement of stimulated emission. Whereas WGM lasing has been subject to intensive studies in micrometer-sized ZnO structures [10], only a small number of publications exists that report on WGM lasing in ZnO nanostructures. This is due to the fact that cavity losses strongly increase when the cross section width d decreases since they are inversely proportional to the structure size.

Nevertheless, Gargas et al. demonstrated WGM lasing in cone-like ZnO structures with diameters ranging 490–920 nm [105], as depicted in Fig. 2.12, but also showed that the WGM lasing threshold strongly increases when the cone diameter drops below 500 nm (see Fig. 2.12g).

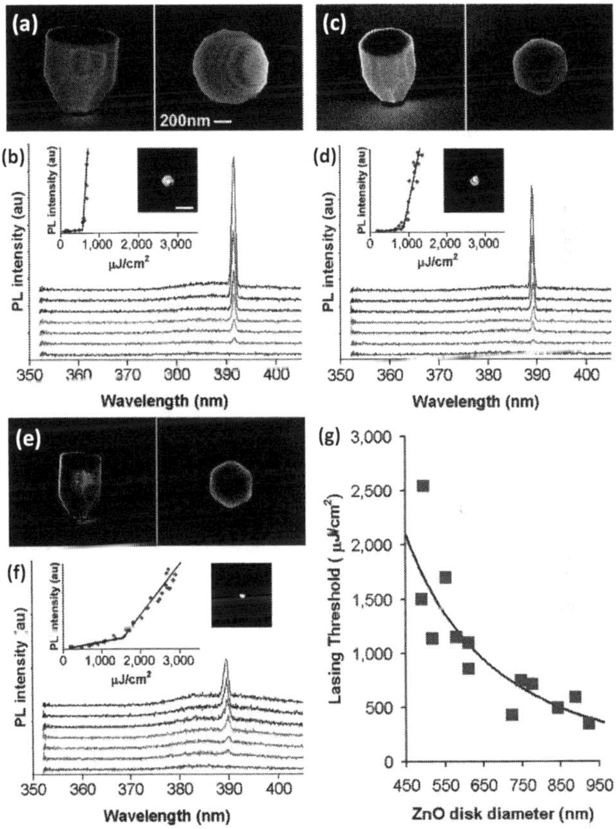

Figure 2.12 SEM pictures of cone-like ZnO nanostructures with $d = 842$ nm (a), 612 nm (c) and 491 nm (e) together with respective excitation-dependent lasing spectra (b, d, f). (g) Lasing threshold vs. diameter. Reprinted from [105] with permission. Copyright © 2010 American Chemical Society.

2.3.2 Fabry–Perot Modes

Besides WGM also Fabry–Perot modes (FPM) can establish in the cross section plane by reflections between parallel facets. However, these light waves are subject to substantial mirror losses due to the low reflectivity of light at the ZnO/air interface, which is caused by the small index contrast for incident light waves. Up to now, this FPM has only been observed in ZnO structures with diameters exceeding 10 μm [106], but should theoretically also be permitted for nanometer-sized ZnO structures. However, the probability to observe them is very low.

More likely, in elongated structures, such as wires or tubes, FPM occur perpendicular to the cross section plane by reflections between the wire end facets in nanowires with constant diameter [107, 108], but have also been demonstrated in tapered nanowires [109, 110]. This is possible since light waves pass several tens of micrometers along the wire axis before they are reflected back, instead of only several nanometers in the cross section. The optical pathway for constructive interference is in this case equal to twice the wire length.

Similar to WGM also FPM can exhibit very high quality factors and thus show lasing action in ZnO nanostructures as long as lasing threshold has been overcome [111]. Huang et al. were in first place to report on FPM lasing phenomena in ZnO nanostructures in 2001 [112]. Since then, numerous reports on stimulated emission in ZnO have been published.

In general, lasing can be driven by different gain mechanisms that are exciton–exciton scattering (EES), electron–hole plasma (EHP), or polariton–polariton scattering (PPS) [6]. All lasing phenomena are accompanied by a strong drop of the mode linewidth (several orders of magnitude) when excitation intensity is increased above lasing threshold.

When the excitation intensity and with that the exciton density is increased only little (but large enough for stimulated emission), EES is more likely to be observed since it occurs due to inelastic collisions between individual excitons. When excitation is even further increased, EHP becomes more probable. It sets in when the exciton Mott density is reached. In this case, bandgap renormalization takes place that red-shifts the EHP emission (in contrast to EES). Instead of excitation-dependent measurements, EHP and EES can more easily be distinguished by their time dependence—an EHP transient decays much faster than an EES transient [45, 113].

Figure 2.13 a shows an optically excited ZnO nanowire below (left) and above lasing threshold (right). As can clearly be seen from

Fig. 2.13a, the wire is governed by spontaneous emission below threshold across the whole wire surface, but exhibits two dot-like light sources of stimulated emission at the wire end facets above threshold. The emission even shows diffraction patterns indicating high spatial coherence [107, 116].

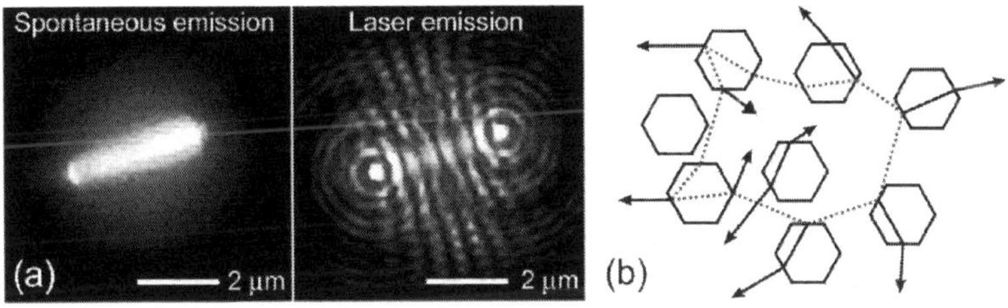

Figure 2.13 Panchromatic photoluminescence images for excitation intensities 24 W/cm^{-2} (left) and 268 W/cm^{-2} (right) of a ZnO nanowire. Reprinted from Ref. [114] with permission. Copyright © 2006 American Chemical Society. (b) Schematic drawing of closed-loop formation in presence of recurrent scattering. Adapted from [115].

A special case of lasing action is stimulated emission of exciton–polaritons by polariton–polariton scattering. Exciton–polaritons are strongly coupled half matter, half light quasiparticles that have bosonic character and can thus undergo Bose–Einstein condensation and polariton lasing by polariton relaxation into the highly occupied bosonic ground state [117]. Polariton lasing is accompanied by extremely low lasing thresholds. Note that PPS leads to a blue-shift of the mode energy due to repulsive interactions between polaritons. Although exciton–polaritons have been observed in ZnO nanowires [118, 119] their unambiguous lasing action has not been demonstrated up to now.

An alternative way to achieve the strong coupling regime between light and cavity has been presented by Schmidt-Grund et al., who fabricated ZnO core–shell nanowires with multi-layered Al_2O_3/YSZ Bragg mirrors as shell layers in order to severely enhance the optical confinement [120], see Fig. 2.14. They achieved Bragg reflectivities of > 90% around the resonator mode and showed anti-crossing behavior of cavity modes and ZnO excitons depending on both temperature and wire thickness. While for planar structures polariton lasing has been reported [121], such effect has not been observed for wires yet.

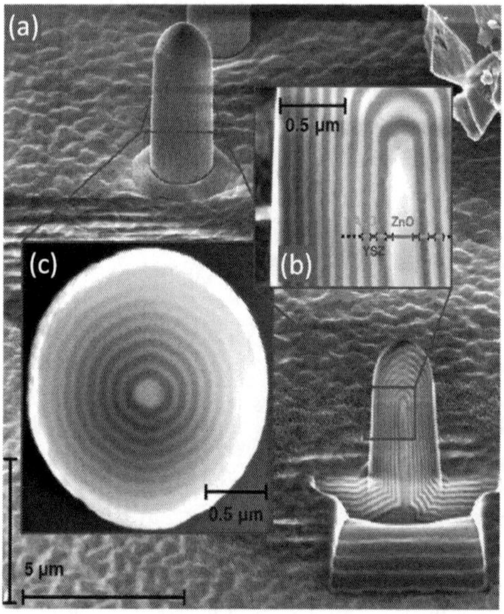

Figure 2.14 SEM image of free standing ZnO NW core cavities surrounded by shells of multilayer Al_2O_3/YSZ Bragg mirrors (a). Inset (b) demonstrates the axial and radial thickness homogeneity of the ZnO core—Al_2O_3/YSZ multilayer shell structure in a plane parallel to the ZnO *c*-axis. Inset (c) shows a cross section perpendicular to the ZnO *c*-axis demonstrating the radial symmetry. Figure is taken from our work [120]. Copyright 2010 Wiley-VCH.

2.3.3 Random Lasing

Closed, optical loop pathways can even develop in disordered dielectric media in the case of strong scattering when recurrent scattering events provide coherent feedback [122]. Optical gain has to be larger than arising losses in order to achieve random lasing. It has been shown that the scattering free mean path has to be reduced to or below the wavelength of the light [123]. Then, the probability that light returns back to the point where it has been scattered before is strongly enhanced [124]. More closed loop paths can be formed.

Experimentally, several criteria have emerged in the last years that need to be fulfilled for the generally accepted realization of a "random laser" in order to distinguish them from conventional lasers. First it has to be proven that lasing modes do not originate from individual nanostructures that act as single resonators, as it is the case for WGM and FPM resonators, see above, rather than originate from several disordered structures that represent strong scattering sources. Further, photoluminescence (PL) experiments with varying laser spot diameters should show a different number of lasing peaks in PL spectra since the change of the spot also affects the number of closed loop paths. And additionally, different lasing spectra need to be detected for different emission angles because arbitrary closed loop paths also exhibit different output directions.

Random lasing was demonstrated in several ZnO nanostructures, such as ZnO powder [124, 125], see Fig. 2.15, ZnO nanowires [115], ZnO nanoparticles [126] or polycrystalline ZnO thin films [123, 127]. Fallert et al. were even able to experimentally show that random laser modes in ZnO powder can be subdivided in either strongly localized or extended modes [125]. The observation of both kinds of modes strongly depends on the spatially present optical gain and thus present mode quality factors.

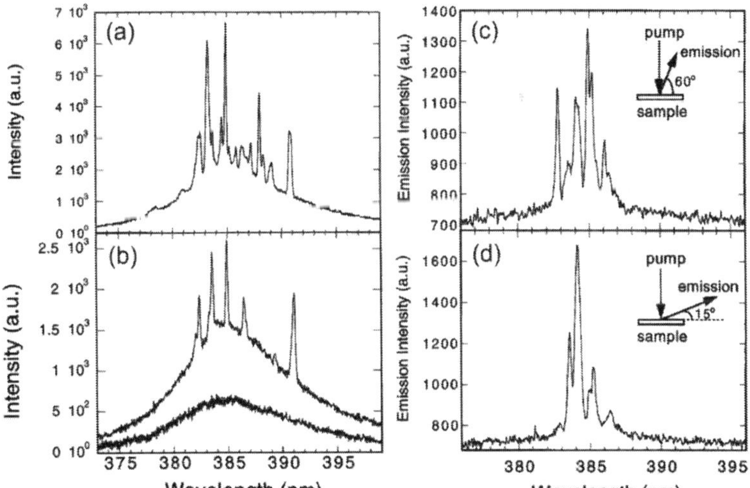

Figure 2.15 Emission spectra of randomly lasing ZnO powder for different excitation areas (a,b) and different detection angles (c,d). The excitation area-dependent spectra were recorded at a fixed excitation intensity of 1012 kW/cm^2 for areas (a) 980 μm^2 and (b) 1870 μm^2. The emission angles are (c) 60° and (d) 15° for a fixed intensity of 1188 kW/cm^2 and an area of 1130 μm^2. Reprinted from [124] with permission. Copyright 1999 by The American Physical Society.

2.4 Raman and Infrared Spectra from Nanostructures

The lattice vibrations in a crystalline solid are directly related to the state of the lattice, particularly to the crystal structure with different kind of defects, and strain. Therefore the investigation of the phonon mode parameters such as phonon energy and broadening as well as (relative) scattering intensities is of high interest to gain access to these material properties. Typically, optical methods as infrared (IR) spectroscopy (reflection, transmission, ellipsometry) and Raman scattering spectroscopy are preferred for studying phonons as they are nondestructive and require little effort compared to, for example, neutron scattering. When it comes to single nanostructures, Raman scattering is superior to the IR techniques since the excitation is carried out using visible or UV light allowing for a much higher spatial resolution. In practice, typically a lateral resolution of 1 μm for micro-Raman setups is achieved which is more than one order of magnitude below the Abbe limit for light in the phonon's energy range.

The wurtzite structure of ZnO under ambient conditions belongs to the point group C_{6v} and consists of four atoms per unit cell, which results in 12 phonon mode branches. According to group theory, the 9 zone center optical phonon modes belong to the irreducible representation

$$\Gamma^{\text{opt}} = A_1 + 2B_1 + E_1 + 2E_2, \tag{2.3}$$

where the modes with E-symmetry are twofold degenerate. The polar A_1 and E_1 modes are both Raman and IR active and split into transversal (TO) and longitudinal optical (LO) modes at the Γ-point. The two E_2 modes are Raman active only, while the two B_1 modes are silent, i.e., they are neither Raman nor IR active.

The Raman spectra of a ZnO single crystal and a ZnO nanobelt [128] for two different crystal orientations are shown in Figs. 2.16 and 2.17, respectively. Essentially, the same peaks are visible for both samples: Besides the Raman active A_1(TO), E_1(TO) and $E_2^{(2)}$ a strong multi-phonon peak near 332 cm^{-1} can be seen which is attributed to the $E_2^{(2)}$–$E_2^{(1)}$ process [129]. The exceptional case of the LO phonons will be discussed below. Differences in relative intensities between single crystals and nanostructures have not been subject of investigations so far. However, the differences between the examples shown here might simply originate from a different laser polarization with respect to the crystal orientation. Polarization dependent Raman measurements of single ZnO nanorods with a diameter of about 200 nm have been carried out by Chien et al. [130] confirming that the selection rules are still valid for their samples.

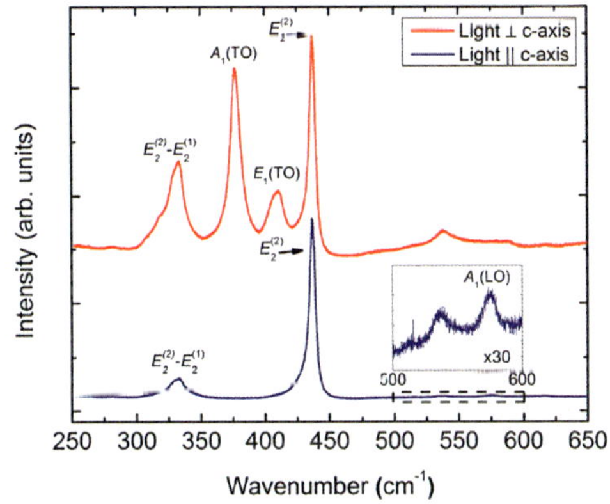

Figure 2.16 Unpolarized Raman spectrum of a ZnO single crystal for backscattering parallel and perpendicular to the *c*-axis. The inset shows the magnified view of the spectrum in the dashed box.

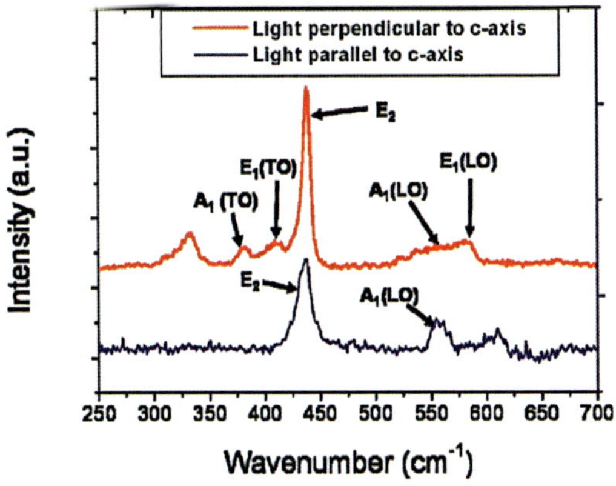

Figure 2.17 Unpolarized Raman spectrum of a ZnO nanobelt for backscattering parallel and perpendicular to the *c*-axis. Reprinted with permission from [128]. Copyright 2009 American Chemical Society.

The phonon mode energies of ZnO nanostructures are compared to those of bulk ZnO in Table 2.3. Only small deviations are observed which do not exceed the variation between different measurements of nanostructures as well as of bulk ZnO. The energetic position of the LO phonons is not given in the table for several reasons. The scattering cross section of both, A_1(LO) and E_1(LO), are much smaller than those of the other modes in case of non-resonant excitation [131] and therefore hardly observed in single crystals and even more so in nanostructures, see also Fig. 2.16.

Table 2.3 Raman shifts in cm^{-1} of main peaks for bulk [129] and nanostructured [130, 133, 134, 137–140] ZnO

	Phonon symmetry				
	$E_2^{(1)}$	$E_2^{(2)}$–$E_2^{(1)}$	A_1(TO)	E_1(TO)	$E_2^{(2)}$
Bulk	99	333	378	410	438
Nanostructures	99	332	380	409	438

Note: Values for nanostructured ZnO are mean values of all publications.

However, in some cases, a peak in the energy range of the LO modes, i.e., 574 to 591 cm^{-1} [129], is detected anyhow. This enhancement of the LO mode intensity can be explained by the introduction of defects [132]—in case of nanostructures this may also be caused by the largely increased surface being also a lattice defect. Excitons bound to such defects provide real electronic states within the band gap highly increasing the scattering cross section by extrinsic Fröhlich interaction.

Since this type of electron–phonon interaction is limited to LO phonons, only these modes undergo an enhancement. The breakdown of translation symmetry in the vicinity of these bound states allows the momentum conservation to be violated. Therefore the selection rules for zone center phonons do not apply anymore and a superposition of contributions from both LO branches is observed (quasi-LO) instead of two separate peaks ([92, 133, 134]). Since the peak position of this quasi-LO can be influenced by defect type and concentration the observed peak positions can hardly be compared.

The effect of the crystal size on Raman spectra has been calculated for semiconductors in general by Campbell and Fauchet [135]. They analyzed the increasing Raman scattering from off-center phonons with decreasing particle size depending on the crystal shape. The phonon dispersion was the only material dependent parameter included in the relation. For phonons with a maximum energy at the Γ-point, as it is typical for optical phonons of semiconductors with two-atomic base, this leads to a red-shift of the Raman peaks accompanied by an asymmetric broadening. In case of ZnO there are also optical phonon branches whose energy increases with increasing k-vector [136] (E_2 and TO modes) which is consequently supposed to lead to a blue-shift of the respective Raman peaks. For silicon, a relative shift of 1 cm^{-1} is predicted for spherical particles of diameters below 15 nm and for cylindrical particles of diameters below 8 nm. Due to the similar absolute variations of phonon mode energy throughout the Brillouin zone, these values can be transferred to ZnO. They are in agreement with the non-significant deviations from bulk ZnO in Table 2.3, because the nanostructures investigated in the given publications had dimensions above this limit.

Phonon confinement and surface (or interface) optical (SO) modes were studied theoretically by Fonoberov and Baladin for crystals with wurtzite structure [141]. They predict confined quasi-TO and -LO modes with energies between those of the corresponding bulk modes with A_1 and E_1 symmetries, which do not depend on the size of the crystal.

Unfortunately, no upper limit of crystal size was predicted above which this effect vanishes so it cannot be stated if it is observable in a certain sample or not. Aside from that, the experimental data cited within the publication to verify the theoretical results are quite questionable and cannot confirm them beyond doubt. The energies of the SO modes were found to lie in between those of the TO and LO modes and to decrease with increasing dielectric constant of the surrounding medium. This is in qualitative accordance with the experimental results of Zeng et al. for spherical Zn/ZnO core–shell nanostructures [143]. A quantitative comparison of these results is complicated by the fact that the SO mode energy was found to be also dependent on the thickness of the ZnO shell. Several other authors claim to have observed such SO modes for pure ZnO nanostructures [92, 144]; the former also by means of IR transmission spectroscopy. However, these authors did not verify this assumption by changing the dielectric constant of the surrounding medium of their samples.

As explained above, IR spectroscopy on the nanoscale is a difficult topic. However, some of the rare published IR spectra of nanowire ensembles are depicted in Fig. 2.18 [142]. The main difference between the nanowires and the thin film is the contribution of the out-of-plane A_1-symmetry phonon mode to the spectra probably caused by non-perfect alignment also of the nominally aligned nanowires. In the same publication, also a pseudo-dielectric function of these ensembles was obtained by fitting Lorentzian oscillators to the experimental data. Again, the main deviation between thin film and nanowire ensembles was the contribution of the phonon modes with A_1 symmetry.

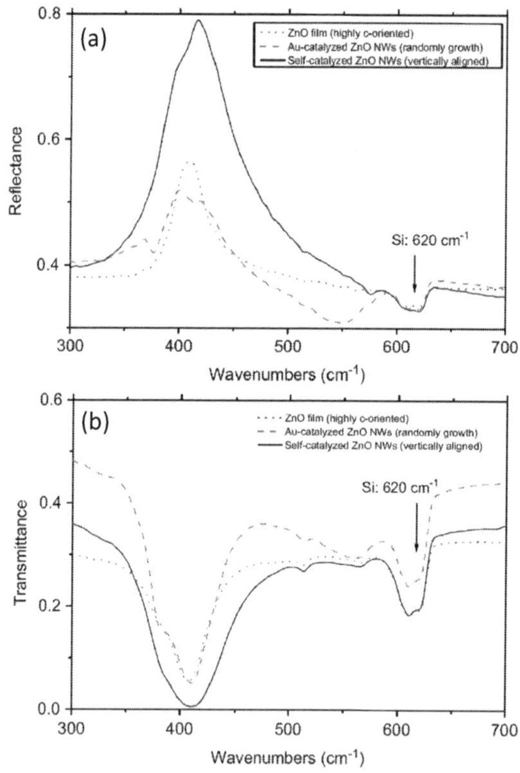

Figure 2.18 Reflection (a) and transmission (b) spectra of randomly and vertically aligned ZnO nanowires in comparison with a *c*-axis oriented thin film. All samples were grown on silicon substrate, the direction of light propagation was normal to the sample surface. Reprinted from [142], with permission from Elsevier.

2.5 Nanostructure Devices

ZnO nanostructures were applied as electrically pumped light-emitting diodes and lasers, as wavelength selective photodetectors, and finally for second harmonics generation.

2.5.1 Light-Emitting Diodes and Lasers

Electrically pumped ZnO-based LEDs and lasers require ZnO-based p–n junctions, which are in general still a rather challenging task. The ZnO community still suffers from a reliable and long-term stable p-type conducting ZnO material, both in form of thin films and nanowires. A general breakthrough in this direction has not been achieved up to now. For thin films, recent attempts to get ZnO p–n junctions and LEDs include ZnO codoped with tellurium and nitrogen [145]. In 2005 the group of Masashi Kawasaki presented the first p-type ZnO:N/i-ZnO/n-ZnO homojunction thin film LED [146, 147]. However, the electroluminescence of this structure was centered at about 440 nm optical wavelength, which is slightly above the bandgap-related emission. Electrically pumped UV lasing was reported for a ZnMgO/ZnO quantum well embedded in a Sb-doped p-type ZnO–Ga-doped n-type ZnO junction [148]. The lasing feedback was provided by close-loop scattering from closely packed nanocolumnar ZnO grains, and the output power was 0.5 μW [148]. The UV-electroluminescence from an n-ZnO:Ga/p-ZnO:Sb LED structure was later reproduced with an output power of 32 nW [149]. A remarkable progress toward higher output power was recently reported by the Kawasaki group. They achieved 70 μW integrated output (from 350 to 450 nm) with an n-ZnO/p-ZnMgO:N thin film UV-LED, which is (only) about one order of magnitude below an InGaN LED under same injection conditions [150].

Optically pumped emission from ZnO nanostructures with increasing appearance of sharp resonance lines with increasing excitation power has already been discussed in Section 2.3.3. Here, we concentrate on electrically pumped emission from ZnO nanowire structures. As mentioned above for ZnO-based thin film p–n junctions, the growth of nanowire p–n junctions is a rather challenging task. Phosphorous-doped ZnO nano- and microwires with diameter below about 5 μm showed p-type behavior for at least 3 to 6 months in bottom-gate field-effect transistors and p-n-diodes, see [152, 153] and references therein. However, recently much thicker dodecagonal ZnO:P microwires with diameter of about 30 μm did not show any p-type conductivity in Hall effect measurements [154]. Because of these difficulties to obtain long term-stable p–n junctions with ZnO nanowires, alternative hybrid LED structures based on n-type ZnO wires and p-type conducting crystalline or amorphous substrate are proposed, see the review [12].

As examples for such hybrid LED structures, n-type ZnO nanowires were prepared on heavily p-doped Si wafers [155], or p-type ZnO:P nanonails were grown on n-type Si substrates [156]. While the n-ZnO/p-Si structure shows electroluminescence in excellent correlation to low-temperature photoluminescence with excitonic origin [155], the p-ZnO:P/n-Si structure shows electrically driven UV lasing peaks around 390 nm wavelength. Only recently, electrically pumped, bright laser emission at room temperature was obtained from a p-type ZnO:Sb nanowire/n-type ZnO thin film structure on c-plane sapphire, see Fig. 2.19.

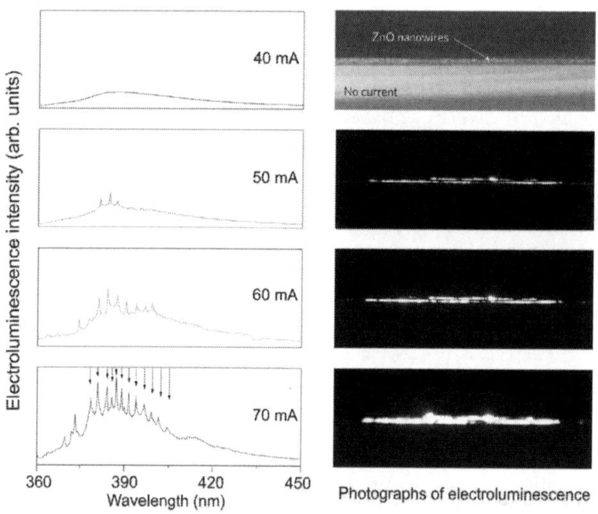

Figure 2.19 (Left) Electroluminescence spectra of a p-type ZnO:Sb NW array on n-type ZnO film with the indicated injection currents. Above 40 mA, Fabry–Perot resonance peaks are visible (see arrows). (Right) Photographs of the laser emission taken with an optical microscope, corresponding to the spectra at the *left* side. The *top right* image is taken with external illumination and without injection current. A superlinear intensity-current dependence is visible. Adapted from [151].

The spectral emission maximum appears at 385 nm and Fabry–Perot waveguide lasing was identified as the dominating emission mechanism [151]. As the major precondition for lasing, the modal gain values of ZnO micro- and nanowires were experimentally determined to be as large as 4800 cm^{-1} under excitation with fs pulses at 266 nm wavelength, which is much larger than modal gain values of bulk ZnO [157]. The modal gain shows a strong variation with the wire diameter and a maximum value for $d \approx 1.2$ μm, in good agreement with theoretical predictions.

2.5.2 Photodetectors

Due to the large surface-to-volume ratio and the Debye length comparable to their size, ZnO nanowire structures show superior electrical sensitivity to light. Two mechanisms of photoresponse in ZnO have been identified, namely (i) the fast band-to-band recombination with time constant in the nanosecond-range, and (ii) the hole trapping due to chemisorbed oxygen molecules at the NW surface [158, 159]. Mechanism (ii) is considered to be dominant in NW materials and is shown schematically in Fig. 2.20. Ambient oxygen molecules are absorbed at the NW surface and capture free electrons from the n-type ZnO (Fig. 2.20b). As a result, a depletion layer with lower conductivity is formed at the surface. Under UV light illumination, electron–hole pairs are generated in the bulk. The holes are assumed to move to the surface due to the band bending and to discharge the negatively charged absorbed oxygen molecules (Fig. 2.20c). By this, the oxygen is photodesorbed from the NW surface, and as a result of hole trapping at dangling bonds, the surface conductivity greatly increases due to the photon interaction [158].

ZnO-based photodetectors are considered to be visible-blind because a photon energy higher than the band gap energy of 3.37 eV is required for the above described mechanisms. Until 2010, more than 100 research papers are published on photodetector properties of 1D-ZnO NWs, see for example the review [159].

Photodetectors based on NWs with *ohmic* contacts show typical dark currents from 2.5 to 10 nA, and photocurrents under UV illumination which are about 2 to 7700 times higher. The latter number is also considered as photoconductive gain of the device. Current rise times of the photocurrent are typically from 0.5 to several hundred seconds [159, 161, 162]. Decay times are usually higher by a factor of at least two. The quantum efficiency was determined to be 12.6% [161]. The time response could be considerably shortened down to the 0.1 ms range [163], which was attributed to the influence of water vapor, which attracts both electrons and holes at the ZnO surface, in contradiction to electron attraction by oxygen molecules, see Fig. 2.20.

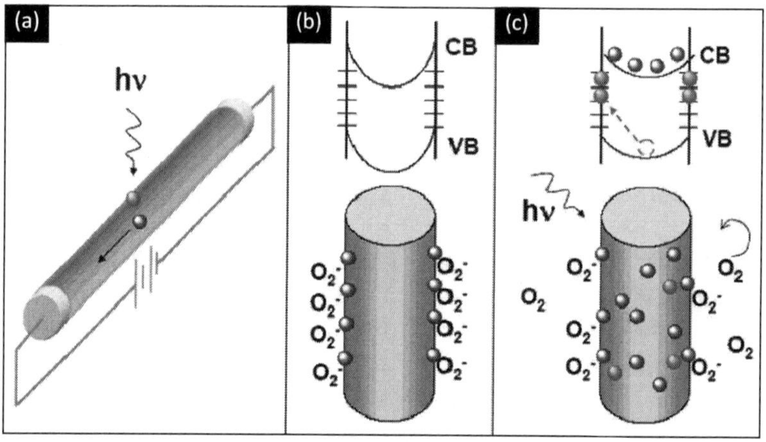

Figure 2.20 Scheme of photoconduction in single NWs. (a) Contact configuration of a NW photodetector. Illumination with photon energy above E_g of the NW material is required for electron–hole pair generation in (a). Under an applied electric field, the unpaired electrons (c) are collected at the anode, which leads to the increase in conductivity. (b, c) Detailed trapping and photoconduction mechanism in ZnO NWs. On top are shown the energy band diagrams of a NW in dark (b), and upon illumination (c). Photogenerated holes migrate to the surface and are trapped in gap states. (c) Oxygen molecules are desorbed from the NW surface. Reprinted from [160] with permission. Copyright 2007 American Chemical Society.

For the competitive surface effects of oxygen and water on UV photoresponse see also [164]. In addition, other surface coatings, such as functional polymers, were found to giantly improve the UV response speed down to the millisecond range [158]. Further insight into the process of capture and release of photogenerated holes by surface trap states was obtained by observation of ultralow-frequency photocurrent oscillations in ZnO nanowires, which was attributed to thickness oscillations of the surface depletion region [165].

Considerable improvements of photodetector characteristics, in particular in terms of dark current, reset time, and internal gain, could be obtained by using *Schottky* barriers [158, 166–171] or *p–n junctions* [172, 173] as the electrical connects to the ZnO NW photosensors. When a reverse-biased Schottky contact is illuminated by 365 nm UV, the photogenerated electron–hole pairs are separated by the strong local electrical field, thus reducing the recombination rate and increasing the carrier lifetime [158]. Furthermore, the Schottky barrier is modified by the oxygen desorption at the ZnO/Pt interface by changing the density of states. These two effects enhance the UV photoresponse of Schottky-type photodetectors. An additional increase of the responsivity of NW photodetectors can be achieved by strain effects, as explained in detail in Fig. 2.21.

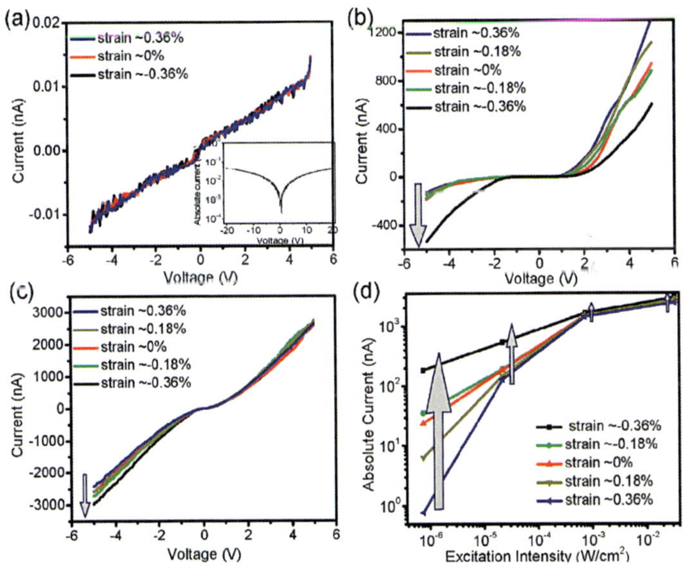

Figure 2.21 Strain effects on ZnO NW photodetectors (a) Typical dark I-V characteristics. (b) I-V curves of the device under different strain with excitation light intensity of 2.2×10^{-5} W/cm^2; responsivity was increased by 190% under-0.36% compressive strain. (c) I-V curves of the device under different strain with excitation light intensity of 3.3×10^{-2} W/cm^2. (d) Absolute photocurrent relative to excitation intensity under different strain. Reprinted from [170] with permission. Copyright 2010 American Chemical Society.

A combined sensitivity of ZnO NW photodetectors to either UV or IR light was achieved in heterostructures with silicon [174–176]. Simultaneous deep-UV and near-IR spectral sensitivity was obtained in n-ZnO/p-Si NW structures [176]. In contrast, a selection of the sensitive spectral range by the polarity of the applied voltage was clearly demonstrated for both n-ZnO NW/n-Si-wafer structure [174], and for p-Si NW core/n-ZnO shell structures [175].

Furthermore, photoresponse effects in ZnO nanoparticles for wavelength from 280 to 375 nm were used to control the surface acoustic wave (SAW) propagation. The SAW phase response exhibits a maximum at 345 nm with fractional acoustoelectronic SAW velocity change per unit power density of 2.8 ppm/(μW/cm^2) [177]. As an extension of the ZnO NW photodetectors toward sensitivities in the deeper UV, Zn_2GeO_4 NWs were successfully used as the photosensitive material [178, 179]. Zn_2GeO_4 has a bandgap energy of 4.68 eV, and is therefore not only visible, but also UV-A/B blind, and only responsive to the UV-C band [179].

2.5.3 Second Harmonic Generation

ZnO is known for its efficient second-harmonic generation (SHG) due to its strong second-order nonlinear susceptibility. In comparison to μm-thick bulk crystals, nanometer thin nonlinear optical films and nanostructures show application-related advantages, for example for optical probing of ultrafast dynamics and broadband nonlinear conversion. From measurements of the angular dependence of the second-harmonic intensity in nanocrystalline ZnO thin films was deduced that bulk effects dominate over surface effects for SHG [180]. A-axis oriented nanocrystallites on ZnO films were found to improve the SHG considerably, in comparison to previous ZnO films without nanocrystallites. Such devices are suitable for advanced pulse characterization techniques [181]. Time-resolved SHG was used to probe both radiative and nonradiative decay in highly excited single ZnO nanowires and nanoribbons [182]. In two-photon absorption measurements on single 80 nm thin ZnO nanowires using 700 to 800 nm excitation, the intensity of the frequency-doubled emission was found to be resonantly enhanced by tuning its wavelength toward the excitonic wavelength of the nanowire around 385 nm [183]. Two-photon absorption was further confirmed to be the main mechanism for SHG in ZnO nanowires by the nearly quadratic dependence of SHG intensity on excitation pulse energy [184]. Stimulated emission from ZnO microtubes with nanometer-thin walls via 808 nm excitation at high fluence above 100 mJ/cm^2 yields sharp excitonic emission peaks with 0.5 to 0.7 nm linewidth, in addition to the weaker SHG peak [185]. A low threshold of 12 mJ/cm^2 for the amplified spontaneous excitonic emission from ZnO nanorod arrays was found upon 800 nm excitation [186].

The SHG in *c*-axis oriented ZnO nanorod arrays (rod diameter 56 to 95 nm) was compared with that in beta-barium borate (type-I BBO) crystals. The maximum effective nonlinearity of ZnO nanowires was 7.5 times higher. Maximum SHG was found for a tilt in between 25° and 45° perpendicularly to the polarization direction of the fundamental excitation, depending on the particular sample [188]. Second harmonic whispering gallery modes (WGM) were observed in tapered legs with hexagonal cross section of ZnO nano-tetrapods, see Fig. 2.22. The second harmonic WGMs in ZnO nanostructures can be very useful for nanophotonic devices such as UV emitters, nanosensors, and wavelength conversion for nanoscale optical circuitry [187].

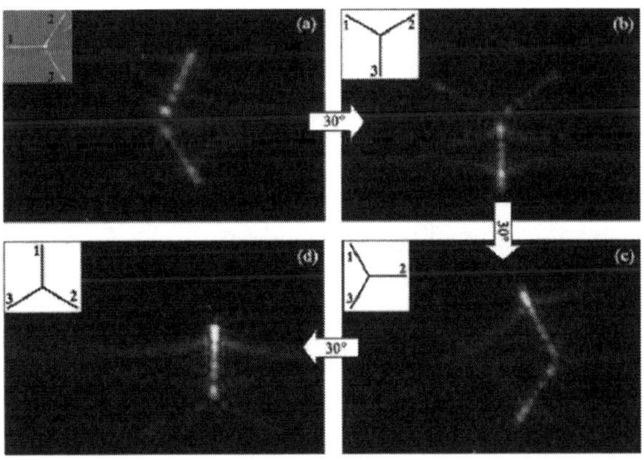

Figure 2.22 Second harmonic images of a ZnO nanowire tetrapod with different orientations relative to the polarization of the fundamental excitation at 810 nm. The polarization is kept vertical in the figures. The tetrapod is rotated 30° clockwise per step from (a) to (d), see insets. Reprinted from [187] with permission. Copyright 2009 American Chemical Society.

SHG can be used also as a tool to monitor structure formation processes. It was observed during formation of laser-induced periodic surface structures, i.e., nanometer-wide ripples, on ZnO surface by intense irradiation at 790 nm [189].

The group of T. Yao in Sendai investigated SHG in polarity controlled ZnO in dependence on grain size and surface roughness [190]. They found strongly increasing nonlinear optical coefficient d_{eff} with decreasing grain size and for surface roughness R_{rms} below 2 nm. Zn-polar ZnO showed higher d_{eff} values as compared to O-polar ZnO. Additionally was found that implantation of H and Zn into ZnO single crystals results in a convinced increment of SHG, while Cu implantation does not [191].

Recently was shown that ultrabroadband excitation of ZnO nanowires with extremely short pulses and high repetition rate of 75.3 MHz enables sufficiently high SHG signals for use in practical applications such as for photodynamic therapy [192]. In the preliminary investigations was found that more than 55% of the energy of a 7 fs pulse, with central frequency at ~800 nm and a FWHM of ~150 nm, from a suitable oscillator can facilitate highly efficient two-photon absorption.

2.6 Summary and Outlook

This chapter provides an overview on recent research on optical properties of and optical devices from ZnO-based nanostructures, including first optical recombination, bandgap engineering and nanoheterostructures. Second, we reviewed cavity effects including whispering gallery and Fabry–Perot modes, regular and random lasing. Furthermore, the current state of Raman and infrared spectroscopy is shown. Finally, optical devices based on ZnO nanowires are demonstrated, i.e., electrically pumped light emitters, photodetectors and nonlinear optical devices. After finishing the writing of this chapter we found another, very recent and application-related review on nanophotonic devices based on ZnO nanowires, edited by W. Zhou and Z. L. Wang, see Ref. [4]. In addition to the optical applications discussed in this chapter, further technological potential for ZnO nanostructures exists for example in the fields of optical interconnects and dye-sensitized solar cells [4]. Because ZnO is a piezoelectric material, the coupling of optical, mechanical, and electrical properties of ZnO nanowires can enable new functional devices. With that, the performance of optoelectronic devices such as photocells and photodetectors can be improved and an effective method to integrate optomechanical devices with microelectronic systems will be provided [4].

Although remarkable progress for optical application of ZnO nanostructures has been achieved in the past, further intense efforts are required to improve the understanding and the control of intrinsic and extrinsic electrically active defects and optical recombination lines of ZnO. Another issue, being particularly important for nanostructures with their high surface-to-volume ratio, is the control and passivation of surfaces and interfaces. As we have shown using the example of the surface-bound excitonic luminescence, surfaces and interfaces may affect significantly the electrical transport properties and optical recombination. Only a combined and simultaneous research work on fundamental materials issues and innovative device concepts will result in improved better device performance and new application directions.

Acknowledgements

The authors acknowledge the financial support by the Deutsche Forschungsgemeinschaft within FOR 1616 "Dynamics and Interactions of Semiconductor Nanowires for Optoelectronics," the European Union within NoE SANDiE NMP4-CT-2004-500101, and the Leipzig Graduate School of Natural Sciences—BuildMoNa (GS 185).

References

1. Wang, Z.L. (2006). Novel nanostructures and nanodevices of ZnO, in *Zinc Oxide Bulk, Thin Films and Nanostructures* (Jagadish, C., and Pearton, S.J., eds.) Chapter 10, Elsevier, Amsterdam, 339–370.
2. Schmidt-Mende, L., and MacManus-Driscoll, J.L. (2007). ZnO-nanostructures, defects, and devices, *Materials Today*, **10,** 40–48.
3. Lorenz, M., Rahm, A., Cao, B.Q., Zuniga-Perez, J., Kaidashev, E.M., Zhakarov, N., Wagner, G., Nobis, Th., Czekalla, C., Zimmermann, G., and Grundmann, M. (2010). Self-organized growth of ZnO-based nano- and microstructures, *Phys. Stat. Sol. B*, **247**, 1265–1281.
4. Yang, Q., Tong, L., Wang, Z.L. (2011). Nanophotonic devices based on ZnO nanowires, *Three-Dimensional Nanoarchitectures* (Zhou, W., and Wang, Z.L., eds.) Chapter 12, Springer, New York, 317–362.

5. Lorenz, M., Lenzner, J., Kaidashev, E.M., Hochmuth, H., and Grundmann, M. (2004). Cathodoluminescence of selected single ZnO nanowires on sapphire, *Ann. Phys. (Leipzig)*, **13**, 39–42.

6. Klingshirn, C., ZnO: From basics towards applications (2007). *Phys. Stat. Sol. B*, **244**, 3027–3073.

7. Park, W.I., An, S.J., Yang, J.L., Yi, G.-C., Hong, S., Joo, T., and Kim, M. (2004). Photoluminescent properties of ZnO/$Zn_{0.8}Mg_{0.2}O$ nanorod single-quantum-well structures, *J. Phys. Chem. B*, **108**, 15457–15460.

8. Nobis, T., Kaidashev, E.M., Rahm, A., Lorenz, M., and Grundmann, M. (2004). Whispering gallery modes in nanosized dielectric resonators with hexagonal cross section, *Phys. Rev. Lett.*, **93**, 103903.

9. Nobis, T., and Grundmann, M. (2005). Low-order optical whispering gallery modes in hexagonal nanocavities, *Phys. Rev. A*, **72**, 063806.

10. Czekalla, C., Sturm, C., Schmidt-Grund, R., Cao, B.Q., Lorenz, M., and Grundmann, M. (2008). Whispering gallery mode lasing in zinc oxide microwires, *Appl. Phys. Lett.*, **92**, 241102.

11. Czekalla, C., Nobis, T., Rahm, A., Cao, B.Q., Zúñiga-Pérez, J., Sturm, C., Schmidt-Grund, R., Lorenz, M., and Grundmann, M. (2010). Whispering gallery modes in ZnO nano- and microwires, *Phys. Stat. Sol. B*, **247**, 1282–1293.

12. Willander, M., Nur, O., Zhao, Q.X., Yang, L.L., Lorenz, M., Cao, B.Q., Zúñiga-Pérez, J., Czekalla, C., Zimmermann, G., Grundmann, M., Bakin, A., Behrends, A., Al-Suleiman, M., El-Shaer, A., Che Mofor, A., Postels, B., Waag, A., Boukos, N., Travlos, A., Kwack, H.S., Guinard, J., and Le Si Dang, D. (2009). Zinc oxide nanorod based photonic devices: recent progress in growth, light emitting diodes and lasers, *Nanotechnology*, **20**, 332001.

13. Volk, J., Håkansson, A., Miyazaki, H.T., Nagata, T., Shimizu, J., and Chikyow, T. (2008). Fully engineered homoepitaxial zinc oxide nanopillar array for near-surface light wave manipulation, *Appl. Phys. Lett.*, **92**, 183114.

14. Lee, B.Y., Sung, M.G., Lee, H., Namgung, S., Park, S.Y., Choi, D.S., and Hong, S. (2010). Integrated devices based on networks of nanotubes and nanowires, *NPG Asia Mater.*, **2**, 103–111.

15. Xu, F., Shen, Y., Sun, L., Zeng, H., and Lu, Y. (2011). Enhanced photocatalytic activity of hierarchical ZnO nanoplate-nanowire architecture as environmentally safe and facilely recyclable photocatalyst, *Nanoscale*, **3**, 5020–5025.

16. Meyer, B.K., Sann, J., Hofmann, D.M., Neumann, C., and A. Zeuner (2005). Shallow donors and acceptors in ZnO, *Semicond. Sci. Technol.*, **20**, 62–66.

17. Meyer, B.K., Sann, J., Eisermann, S., Lautenschläger, S., Wagner, M.R., Kaiser, M., Callsen, G., Reparaz, J.S., and A. Hoffmann (2010). Excited state properties of donor bound excitons in ZnO, *Phys. Rev. B*, **82**, 115207.

18. Meyer, B.K., Sann, J., Lautenschläger, S., Wagner, M.R., and Hoffmann, A. (2007). Ionized and neutral donor-bound excitons in ZnO, *Phys. Rev. B*, **76**, 184120.

19. Wagner, M.R., Callsen, G., Reparaz, J.S., Schulze, J.-H., Kirste, R., Cobet, M., Ostapenko, I.A., Rodt, S., Nenstiel, C., Kaiser, M., Hoffmann, A., Rodina, A.V., Phillips, M.R., Lautenschläger, S., Eisermann, S., and Meyer, B.K. (2011). Bound excitons in ZnO: Structural defect complexes versus shallow impurity centers, *Phys. Rev. B*, **84**, 035313.

20. Meyer, B.K., Alves, H., Hofmann, D.M., Kriegseis, W., Forster, D., Bertram, F., Christen, J., Hoffmann, A., Straßburg, M., Dworzak, M., Haboeck, U., and Rodina, A.V. (2004). Bound exciton and donor acceptor pair recombinations in ZnO, *Phys. Stat. Sol. B*, **241**, 231–260.

21. Huang, K., and Rhys A. (1950), Theory of light absorption and non-radiative transitions in F-centers, *Proc. R. Soc. A*, **204**, 406–423.

22. Hopfield, J. (1959). A theory of edge-emission phenomena in CdS, ZnS and ZnO, *J. Phys. Chem. Solids*, **10**, 110–119.

23. Conradt, J., Sartor, J., Thiele, C., Maier-Flaig, F., Fallert, J., Kalt, H., Schneider, R., Fotouhi, M., Pfundstein, Zibat, P.V., and Gerthsen, D. (2011). Catalyst-free growth of zinc oxide nanorod arrays on sputtered aluminum-doped zinc oxide for photovoltaic applications, *J. Phys. Chem. C*, **115**, 3539–3543.

24. Park, W.I., Jun, Y.H., Jung, S.W., and Yi, G.-C. (2003). Excitonic emissions observed in ZnO single crystal nanorods, *Appl. Phys. Lett.*, **82**, 964–966.

25. Ozaki, S., Tsuchiya, T., Inokuchi, Y., and Adachiy, S. (2005). Photoluminescence and photomodulated transmittance spectroscopy of ZnO nanowires, *Phys. Stat. Sol. A*, **202**, 1325–1335.

26. Kwok, W., Djurišić, A., Leung, Y., Chan, W., Phillips, D., Chen, H., Wu, C., Gwo, S., and Xie, M. (2005). Study of excitonic emission in highly faceted ZnO rods, *Chem. Phys. Lett.*, **412**, 141–144.

27. Tam, K.H., Cheung, C.K., Leung, Y.H., Djurišić, A.B., Ling, C.C., Beling, C.D., Fung, S., Kwok, W.M., Chan, W.K., Phillips, D.L., Ding, L., and Ge, W.K. (2006). Defects in ZnO nanorods prepared by a hydrothermal method, *J. Phys. Chem. B*, **110**, 20865–20871.

28. Robin, I.C., Gauron, B., Ferret, P., Tavares, C., Feuillet, G., Le Si Dang, D., Gayral, B., and Gérard, J.M. (2007). Evidence for low density of nonradiative defects in ZnO nanowires grown by metal organic vapor-phase epitaxy, *Appl. Phys. Lett.*, **91**, 143120.

29. Xia, J. B. and Zhang, X. W. (2006). Electronic structure of ZnO wurtzite quantum wires, *Eur. Phys. J. B*, **49**, pp. 415–420.

30. Xiang, B., Wang, P., Zhang, X., Dayeh, S.A., Aplin, D.P.R., Soci, C., Yu, D., and Wang, D. (2007). Rational synthesis of p-type zinc oxide nanowire arrays Using simple chemical vapor deposition, *Nano Lett.*, **7**, 323–328.

31. Nobis, T., Kaidashev, E.M., Rahm, A., Lorenz, M., Lenzner, J., and Grundmann, M. (2004). Spatially inhomogeneous impurity distribution in ZnO micropillars, *Nano Lett.*, **4**, 797–800.

32. Wischmeier, L., Voss, T., Rückmann, I., Gutowski, J., Mofor, A. C., Bakin, A., and Waag, A. (2006). Dynamics of surface-excitonic emission in ZnO nanowires, *Phys. Rev. B*, **74**, 195333.

33. Grabowska, J., Meaney, A., Nanda, K.K., Mosnier, J.-P., Henry, M.O., Duclère, J.-R., and McGlynn, E. (2005). Surface excitonic emission and quenching effects in ZnO nanowire/nanowall systems: Limiting effects on device potential, *Phys. Rev. B*, **71**, 115439.

34. Stichtenoth, D., Ronning, C., Niermann, T., Wischmeier, L., Voss, T., Chien, C.-J., Chang, P.-C., and Lu, J.G. (2007). Optical size effects in ultrathin ZnO nanowires, *Nanotechnology*, **18**, 435701.

35. Wischmeier, L., Voss, T., Rückmann, I., and Gutowski, J. (2008). Correlations between surface-excitonic emission bands in ZnO nanowires, *Nanotechnology*, **19**, 135705.

36. Richters, J.-P., Voss, T., Kim, D.S., Scholz, R., and Zacharias, M. (2008). Enhanced surface-excitonic emission in ZnO/Al_2O_3 core–shell nanowires, *Nanotechnology*, **19**, 305202.

37. Lin, S., Ye, Z., He, H., Zhao, B., Zhu, L., and Huang, J. (2008). Photoluminescence properties of ZnO nanoneedles grown by metal organic chemical vapor deposition, *J. Appl. Phys.*, **104**, 064311.

38. Biswas, M., Jung, Y.S., Kim, H.K., Kumar, K., Hughes, G.J., Newcomb, S., Henry, M.O., and McGlynn, E. (2011). Microscopic origins of the surface exciton photoluminescence peak in ZnO nanostructures, *Phys. Rev. B*, **83**, 235320.

39. Voss, T., Richters, J.-P., and Dev, A. (2010). Surface effects and nonlinear optical properties of ZnO nanowires, *Phys. Stat. Sol. B*, **247**, 2476–2487.

40. Li, J., and Wang, L.-W. (2005). Band-structure-corrected local density approximation study of semiconductor quantum dots and wires, *Phys. Rev. B*, **72**, 125325.

41. Tuoc, V.N. (2010). The self-consistent charge density functional tight binding study on wurtzite nanowire, *Comput. Mater. Sci.*, **49**, S161–S169.

42. Wang, X., Ding, Y., Summers, C.J., and Wang, Z.L. (2004). Large-scale synthesis of six-nanometer-wide ZnO nanobelts, *J. Phys. Chem. B*, **108**, 8773–8777.

43. Greyson, E., Babayan, Y., and Odom, T. (2004). Directed growth of ordered arrays of small-diameter ZnO nanowires, *Adv. Mater.*, **16**, 1348–1352.

44. Park, W.I., Yoo, J., and Yi, G.-C. (2005). Catalyst-free metalorganic chemical-vapor deposition of ultrafine ZnO nanorods, *J. Korean Phys. Soc.*, **46**, L1067–L1070.

45. Djurišić, A.B., and Leung, Y.H. (2006). Optical properties of ZnO nanostructures, *Small*, **2**, 944–961.

46. Ko, H.J., Chen, Y., Hong, S.K., and Yao, T. (2000). MBE growth of high-quality ZnO films on epi-GaN, *J. Crystal Growth*, **209**, 816–821.

47. Kisi, E.H., and Elcombe, M.M. (1989). u parameters for the wurtzite structure of ZnS and ZnO using powder neutron diffraction, *Acta Cryst.*, **45**, 1867–1870.

48. Roessler, D.M., and Walker, W.C. (1967). Electronic spectrum and ultraviolet optical properties of crystalline MgO, *Phys. Rev.*, **159**, 733–738.

49. Schmahl, N., and Eikerling, G. (1968). Über kryptomodifkationen des Cu(II)-oxids, *Z. Phys. Chem.*, **62**, 268–279.

50. Roessler, D., Walker, W., and Loh, E. (1969). Electronic spectrum of crystalline beryllium oxide, *J. Phys. Chem. Solids*, **30**, 157–167.

51. Bellamy, B., Baker, T., and Livey, D. (1962). The lattice parameter and density of beryllium oxide determined by precise X-ray methods, *J. Nucl. Mater.*, **6**, 1–4.

52. Choi, Y.S., Lee, C.G., and Cho, S.M. (1996). Transparent conducting ZnCdO thin films prepared by the sol-gel process, *Thin Solid Films*, **289**, 153–158.

53. Cimino, A., and Marezio, M. (1960). Lattice parameter and defect structure of cadmium oxide containing foreign atoms, *J. Phys. Chem. Solids*, **17**, 57–64.

54. Ohtomo, A., Kawasaki, M., Koida, T., Masubuchi, K., Koinuma, H., Sakurai, Y., Yoshida, Y., Yasuda, T., and Segawa, Y. (1998). $Mg_xZn_{1-x}O$ as a II–VI widegap semiconductor alloy, *Appl. Phys. Lett.*, **72**, 2466–2468.

55. Takagi, T., Tanaka, H., Fujita, S., and Fujita, S. (2003). Molecular beam epitaxy of high magnesium content single-phase wurzite MgZnO alloys ($x \simeq 0.5$) and their application to solar-blind region photodetectors, *Jpn. J. Appl. Phys.*, **42**, L401–L403.

56. Sadofev, S., Blumstengel, S., Cui, J., Puls, J., Rogaschewski, S., Schafer, P., Sadofyev, Y.G., and Henneberger, F. (2005). Growth of high-quality ZnMgO epilayers and ZnO/ZnMgO quantum well structures by radical-source molecular-beam epitaxy on sapphire, *Appl. Phys. Lett.*, **87**, 091903.

57. Ryu, Y.R., Lee, T.S., Lubguban, J.A., Corman, A.B., White, H.W., Leem, J. H., Han, M.S., Park, Y.S., Youn, C.J., and Kim, W.J. (2006). Wide-band gap oxide alloy: BeZnO, *Appl. Phys. Lett.*, **88**, 052103.

58. Han, M., Kim, J., Jeong, T., Park, J., Youn, C., Leem, J., and Ryu, Y. (2007). Growth and optical properties of epitaxial $Be_xZn_{1-x}O$ alloy films, *J. Cryst. Growth*, **303**, 506–509.

59. Yamamoto, K., Tsuboi, T., Ohashi, T., Tawara, T., Gotoh, H., Nakamura, A., and Temmyo, J. (2010). Structural and optical properties of Zn(Mg,Cd)O alloy films grown by remote-plasma-enhanced MOCVD, *J. Cryst. Growth*, **312**, 1703–1708.

60. Heo, Y.W., Kaufman, M., Pruessner, K., Norton, D.P., Ren, F., Chisholm, M.F., and Fleming, P.H. (2003). Optical properties of $Zn_{1-x}Mg_xO$ nanorods using catalysis-driven molecular beam epitaxy, *Solid-State Electron.*, **47**, 2269–2273.

61. Kling, R., Kirchner, C., Gruber, T., Reuss, F., and Waag, A. (2004). Analysis of ZnO and ZnMgO nanopillars grown by self-organization, *Nanotechnology*, **15**, 1043–1046.

62. Ku, C.-H., Chiang, H.-H., and Wu, J.-J. (2005). Bandgap engineering of well-aligned $Zn_{1-x}Mg_xO$ nanorods grown by metalorganic chemical vapor deposition, *Chem. Phys. Lett.*, **404**, 132–135.

63. Kar, J., Jeong, M., Lee, W., and Myoung, J. (2008). Fabrication and characterization of vertically aligned ZnMgO/ZnO nanowire arrays, *Mater. Sci. Eng. B*, **147**, 74–78.

64. Lorenz, M., Kaidashev, E.M., Rahm, A., Nobis, T., Lenzner, J., Wagner, G., Spemann, D., Hochmuth, H., and Grundmann, M. (2005). $Mg_xZn_{1-x}O$

($0 \leq x < 0.2$) nanowire arrays on sapphire grown by high-pressure pulsed-laser deposition, *Appl. Phys. Lett.*, **86**, 143113.

65. Zhi, M., Zhu, L., Ye, Z., Wang, F., and Zhao, B. (2005). Preparation and properties of ternary ZnMgO nanowires, *J. Phys. Chem. B*, **109**, 23930–23934.
66. Lu, J.G., Zhang, Y.Z., Ye, Z.Z., Zeng, Y.J., Huang, J.Y., and Wang, L. (2007). Rational synthesis and tunable optical properties of quasialigned $Zn_{1-x}Mg_xO$ nanorods, *Appl. Phys. Lett.*, **91**, 193108.
67. Wang, F., Zhao, C., Liu, B., and Yuan, S. (2009). Synthesis and photoluminescence of quasiarrayed ZnMgO nanorods, *J. Phys. D: Appl. Phys.*, **42**, 115411.
68. Chu, G., Zhinqun, X., and Shijia, Z. (2010). Growth of ZnMgO and ZnCdO nanowires and the application in high emitting devices, *J. Nano Sci. Nanotechnol.*, **10**, 4893–4896.
69. Lee, C.Y., Tseng, T.Y., Li, S.Y., and Lin, P. (2006). Effect of phosphorus dopant on photoluminescence and field-emission characteristics of $Mg_xZn_{1-x}O$ nanowires, *J. Appl. Phys.*, **99**, 024303.
70. Pan, C.-J., Hsu, H.-C., Cheng, H.-M., Wu, C.-Y., and Hsieh, W.-F. (2007). Structural and optical properties of ZnMgO nanostructures formed by Mg in-diffused ZnO nanowires, *J. Solid State Chem.*, **180**, 1188–1192.
71. Wang, F.Z., Ye, Z.Z., Ma, D.W., Zhu, L.P., Zhuge, F., and He, H.P. (2005). Synthesis and characterization of quasi-aligned ZnCdO nanorods, *Appl. Phys. Lett.*, **87**, 143101.
72. Li, G.-R., Zhao, W.-X., Bu, Q., and Tong, Y.-X. (2009). A novel electrochemical deposition route for the preparation of $Zn_{1-x}Cd_xO$ nanorods with controllable optical properties, *Electrochem. Commun.*, **11**, 282–285.
73. Shan, C.X., Liu, Z., Zhang, Z.Z., Shen D.Z., and Hark, S.K. (2006). A simple route to porous ZnO and ZnCdO nanowires, *J. Phys. Chem. B*, **110**, 11176–11179.
74. Tang, H., Kwon, B.J., and Park, J.-Y. (2010). Characterizations of individual ZnMgO nanowires synthesized by a vapor-transport method, *Phys. Stat. Sol. A*, **207**, 2478–2482.
75. Wang, F., Liu, B., Zhang, Z., and Yuan, S. (2009). Synthesis and properties of Cd-doped ZnO nanotubes, *Phys. E*, **41**, 879–82.
76. Wang, X., Li, G., and Wang, Y. (2009). Synthesis and characterization of well aligned Cd-Al codoped Zno nanorods arrays, *Chem. Phys. Lett.*, **469**, 308–312.

77. Björk, M.T., Ohlsson, B.J., Sass, T., Persson, A.I., Thelander, C., Magnusson, M.H., Deppert, K., Wallenberg, L.R., and Samuelson, L. (2002). One-dimensional heterostructures in semiconductor nanowhiskers, *Appl. Phys. Lett.*, **80**, 1058–1060.

78. Sun, Y., Cho, Y.-H., Kim, H.-M., and Kang, T.W. (2005). High efficiency and brightness of blue light emission from dislocation-free InGaN/GaN quantum well nanorod arrays, *Appl. Phys. Lett.*, **87**, 093115.

79. Cao, B.Q., Zúñiga-Pérez, J., Boukos, N., Czekalla, C., Hilmer, H., Lenzner, J., Travlos, A., Lorenz, M., and Grundmann, M. (2009). Homogeneous core/shell ZnO/ZnMgO quantum well heterostructures on vertical ZnO nanowires, *Nanotechnology*, **20**, 305701.

80. Kong, B.H., Kim, D.C., Kim, Y.Y., and Cho, H.K. (2007). Growth of vertical nanorods with a MgZnO ternary phase through thermal evaporation, *J. Korean Phys. Soc.*, **50**, 1701–1705.

81. Kong, B.H., Mohanta, S.K., Kim, Y.Y., and Cho, H.K. (2008). Synthesis and characterization of ZnO/MgZnO heterostructure nanorods by simple two-step evaporation, *Nanotechnology*, **19**, 085607.

82. Jang, E.-S., Bae, J.Y., Yoo, J., Park, W.I., Kim, D.-W., Yi, G.-C., Yatsui, T., and Ohtsu, M. (2006). Quantum confinement effect in ZnO/$Mg_{0.2}Zn_{0.8}O$ multishell nanorod heterostructures, *Appl. Phys. Lett.*, **88**, 023102.

83. Lange, M., Dietrich, C.P., Zúñiga-Pérez, J., von Wenckstern, H., Lorenz, M., and Grundmann, M. (2011). MgZnO/ZnO quantum well nanowire heterostructures with large confinement energies, *J. Vac. Sci. Technol. A*, **29**, 03A104.

84. Agarwal, R. (2008). Heterointerfaces in semiconductor nanowires, *Small*, **4**, 1872–1893.

85. Bae, J.Y., Yoo, J., and Yi, G.-C. (2006). Fabrication and photoluminescent characteristics of ZnO/$Mg_{0.2}Zn_{0.8}O$ coaxial nanorod single quantum well structures, *Appl. Phys. Lett.*, **89**, 173114.

86. Park, W., Yi, G.-C., Kim, M., and Pennycook, S. (2003). Quantum confinement observed in ZnO/ZnMgO nanorod heterostructures, *Adv. Mater.*, **15**, 526–529.

87. Yatsui, T., Lim, J., Ohtsu, M., An, S.J., and Yi, G.-C. (2004). Evaluation of the discrete energy levels of individual ZnO nanorod single-quantum-well structures using nearfield ultraviolet photoluminescence spectroscopy, *Appl. Phys. Lett.*, **85**, 727.

88. Kim, C., Park, W.I., Yi, G.-C., and Kim, M. (2006). Formation and photoluminescent properties of embedded ZnO quantum dots in ZnO/ZnMgO multiple-quantum-wellstructured nanorods, *Appl. Phys. Lett.*, **89**, 113106.

89. Bakin, A., El-Shaer, A., Mofor, A.C., Al-Suleiman, M., Schlenker, E., and Waag, A. (2007). ZnMgO-ZnO quantum wells embedded in ZnO nanopillars: Towards realisation of nano-LEDs, *Phys. Stat. Sol. C*, **4**, 158–161.

90. Zippel, J., Heitsch, S., Stölzel, M., Müller, A., von Wenckstern, H., Benndorf, G., Lorenz, M., Hochmuth, H., and Grundmann, M. (2010). Optical properties of homo- and heteroepitaxial ZnO/$Mg_xZn_{1-x}O$ single quantum wells grown by pulsed-laser deposition, *J. Lumin.*, **130**, 520–526.

91. Cheng, C., Liu, B., Sie, E.J., Zhou, W., Zhang, J., Gong, H., Huan, C.H.A., Sum, T.C., Sun, H., and Fan, H.J. (2010). ZnCdO/ZnO coaxial multiple quantum well nanowire heterostructures and optical properties, *J. Phys. Chem. C*, **114**, 3863–3868.

92. Cheng, B., Sun, W., Jiao, J., Tian, B., Xiao, Y., and Lei, S. (2010). Disorder-induced Raman scattering effects in one-dimensional ZnO nanostructures by incorporation and anisotropic distribution of Dy and Li codopants, *J. Raman Spectrosc.*, **41**, 1221–1226.

93. Czekalla, C., Guinard, J., Hanisch, C., Cao, B., Kaidashev, E., Boukos, N., Travlos, A., Renard, J., Gayral, B., Le Si Dang, D., Lorenz, M., and Grundmann, M. (2008). Spatial fluctuations of optical emission from single ZnO/MgZnO nanowire quantum wells, *Nanotechnology*, **19**, 115202.

94. McCall, S.L., Levi, A.F., Slusher, R.E., Pearton, S.J., and Logan, R.A. (1992). Whispering-gallery mode microdisk lasers, *Appl. Phys. Lett.*, **60**, 289–291.

95. Rayleigh, L. (1912). The Problem of the whispering gallery, *Phil. Mag.*, **6**, 1001–1004.

96. Grundmann, M., and Dietrich, C.P., Whispering gallery modes in deformed hexagonal resonators, *Phys. Stat. Sol. B*, **249**, 871–879 (2012)

97. Zhang, X., Zhang, X., Xu, J., Shan, X., Xu, J., and Yu, D. (2009). Whispering gallery modes in single triangular ZnO nanorods, *Opt. Lett.*, **34**, 2533–2535.

98. Nobis T., Rahm A., Czekalla C., Lorenz M., and Grundmann M. (2007). Optical whispering gallery modes in dodecagonal zinc oxide microcrystals, *Superlatt. Microstr.*, **42**, 333–336.

99. Dong, H.X., Chen, Z.H., Sun, L.X., Xie, W., Tan, H.H., Liu, J., Jagadish, C., and Shen X.C. (2010). Single-crystalline hexagonal ZnO microtube optical resonators, *J. Mater. Chem.*, **20**, 5510–5515.

100. Wiersig, J. (2003). Hexagonal dielectric resonators and microcrystal lasers, *Phys. Rev. A*, **67**, 023807.

101. Yang, Y.H., Zhang, Y., Wang, N.W., Wang, C.X., Li, B.J., and Yang, G.W. (2011). ZnO nanocone: application in fabrication of the smallest whispering gallery optical resonantor, *Nanoscale*, **3**, 592–597.

102. Taflove, A., and Brodwin, M.E. (1975). Numerical solution of steady-state electromagnetic scattering problems using the time-dependent Maxwell's equations, *IEEE Trans. Microw. Theory Tech.*, **23**, 623–630.

103. Pavlovic, G., Malpuech, G., and Gippius, N.A. (2010). Dispersion and polarization conversion of whispering gallery modes in nanowires, *Phys. Rev. B*, **82**, 195328.

104. Schmidt-Grund, R., Kühne, P., Czekalla, C., Schumacher, D., Sturm, C., and Grundmann, M. (2011). Determination of the refractive index of single crystal bulk samples and micro-structures, *Thin Solid Films*, **519**, 2777–2781.

105. Gargas, D.J., Moore, M.C., Ni, A., Chang, S.-W., Zhang, Z.Y., Chuang, S.-L., and Yang P.D. (2010). Whispering gallery mode lasing from zinc oxide hexagonal nanodisks, *ACS Nano*, **4**, 3270–3276.

106. Dietrich, C.P., Lange, M., Sturm, C., Schmidt-Grund, R., and Grundmann M. (2011). One-and two-dimensional cavity modes in ZnO microwires, *New J. Phys.*, **13**, 103021.

107. Voss, T., Svacha, G.T., Mazur, E., Müller, S., Ronning, C., Konjhodzic, D., and Marlow, F. (2007). High-Order waveguide modes in ZnO nanowires, *Nano Lett.*, **7**, 3675–3680.

108. Li, H.-Y., Rühle, S., Khedoe, R., Koenderink, A.F., and Vanmaekelbergh, D. (2009). Polarization, microscopic origin, and mode structure of luminescence and lasing from single ZnO nanowires, *Nano Lett.*, **9**, 3515–3520.

109. Lee, S.H., Goto, T., Miyazaki, H., Chang, J., and Yao, T. (2010). Optical resonant cavity in a nanotaper, *Nano Lett.* **10**, 2038–2042.

110. Xu, X.L., Brossard, F.S.F., Williams, D.A., Collins, D.P., Holmes, M.J., Taylor, R.A., and Zhang, X.T. (2010). Cavity modes of tapered ZnO nanowires, *New J. Phys.*, **12**, 083052.

111. Zimmler, M.A., Bao, J.M., Capasso, F., Müller, S., and Ronning, C. (2008). Laser action in nanowires: observation of the transition from amplified spontaneous emission to laser oscillation, *Appl. Phys. Lett.*, **93**, 051101.

112. Huang, M.H., Mao, S., Feick, H., Yan, H., Wu, Y.Y., Kind, H., Weber, E., Russo, R., and Yang, P.D. (2001). Room-temperature ultraviolet nanowire nanolasers, *Science*, **292**, 1897–1899.

113. Djurišić, A.B., Kwok, W.M., Leung, Y.H., Phillips, D.L., and Chan W.K. (2005). Stimulated emission in ZnO nanostructures: a time-resolved study, *J. Phys. Chem. B*, **109**, 19228–19233.

114. van Vugt, L.K., Rühle, S., and Vanmaekelbergh, D. (2007). Phase-correlated nondirectional laser emission from the end facets of a ZnO nanowire, *Nano Lett.*, **6**, 2707–2711.

115. Yu, S.F., Yuen, C., Lau, S.P., Park, W.I., and Yi, G.-C. (2004). Random laser action in ZnO nanorod arrays embedded in ZnO epilayers, *Appl. Phys. Lett.*, **84**, 3241–3243.

116. Johnson, J.C., Yan, H.Q., Yang, P.D., and Saykally, R.J. (2003). Optical cavity effects in ZnO nanowire lasers and waveguides, *J. Phys. Chem. B*, **107**, 8816–8828.

117. Kasprzak, J., Richard, M., Kundermann, S., Baas, A., Jeambrun, P., Keeling, J.M.J., Marchetti, F.M., Szymańska, M.H., André, R., Staehli, J.L., Savona, V., Littlewood, P.B., Deveaud, B., and Le Si Dang (2006). Bose–Einstein condensation of exciton polaritons, *Nature*, **443**, 409–414.

118. van Vugt, L.K., Rühle, S., Ravindran, P., Gerritsen, H.C., Kuipers, L., and Vanmaekelbergh, D. (2006). Exciton polaritons confined in a ZnO nanowire cavity, *Phys. Rev. Lett.*, **97**, 147401.

119. Rühle, S., van Vugt, L.K., Li, H.-Y., Keizer, N.A., Kuipers, L., and Vanmaekelbergh, D. (2008), Nature of sub-band gap luminescent eigenmodes in a ZnO nanowire, *Nano Lett.*, **8**, 119–123.

120. Schmidt-Grund, R., Hilmer, H., Hinkel, A., Sturm, C., Rheinländer, B., Gottschalch, V., Lange, M., Zúñiga-Pérez, J., and Grundmann, M. (2010). Two-dimensional confined photonic wire resonators—strong light-matter coupling, *Phys. Stat. Sol. B*, **247**, 1351–1364.

121. Franke H., Sturm C., Schmidt-Grund R., Wagner G., and Grundmann M. (2012). Ballistic propagation of exciton–polariton condensates in a ZnO-based microcavity, *New J. Phys.*, **3**, 013037.

122. Wiersma, D.S. (2008). The physics and applications of random lasers, *Nat. Phys.*, **4**, 359–367.

123. Cao, H., Zhao, Y.G., Ong, H.C., Ho, S.T., Dai, J.Y., Wu, J.Y., and Chang, R.P.H. (1998). Ultraviolet lasing in resonators formed by scattering in semiconductor polycrystalline films, *Appl. Phys. Lett.*, **73**, 3656–3658.

124. Cao, H., Zhao, Z.G., Ho, S.T., Seelig, E.W., Wang, Q.W., and Zhang, R.P.H. (1999). Random laser action in semiconductor powder, *Phys. Rev. Lett.*, **82**, 2278–2281.

125. Fallert, J., Dietz, R.J.B., Sartor, J., Schneider, D., Klingshirn, C., and Kalt H. (2009). Co-existence of strongly and weakly localized random laser modes, *Nat. Photon.*, **3**, 279–282.

126. Lau, S.P., Yang, H.Y., Yu, S.F., Yuen, C., Leong, E.S.P., Li, H.D., and Hng, H.H. (2005). Flexible ultraviolet random lasers based on nanoparticles, *Small*, **1**, 956–959.

127. Zhu, H, Shan, C.-X., Zhang, J.-Y., Zhang, Z.-Z., Li, B.-H., Zhao, D.-X., Bin, Y., Shen, D.-Z., Fan, X.-W., Tang, X.-K., Hou, X.H., and Choy, K.-L. (2010). Low-threshold electrically pumped random lasers, *Adv. Mat.*, **22**, 1877–1881.

128. Singamaneni, S., Gupta, M., Yang, R., Tomczak, M.M., Naik, R.R., Wang, Z.L., and Tsuruk, V.V. (2009). Nondestructive in situ identification of crystal orientation of anisotropic ZnO nanostructures, *ACS Nano*, **3**, 2593–2600.

129. Cuscó, R., Alarcón-Lladó, E., Ibáñez, J., Artús, L., Jiménez, J., Wang, B., and Callahan, M.J. (2007). Temperature dependence of Raman scattering in ZnO, *Phys. Rev. B*, **75**, 165202.

130. Chien, C., Wu, M., Chen, C., Yang, H., Wu, J., Su, W., Lin, C., and Chen, Y. (2008). Polarization-dependent confocal Raman microscopy of an individual ZnO nanorod, *Appl. Phys. Lett.*, **92**, 223102.

131. Calleja, J.M., and Cardona, M. (1977). Resonant Raman scattering in ZnO, *Phys. Rev. B*, **16**, 3753–3761.

132. Friedrich, F., and Nickel, N.H. (2007). Resonant Raman scattering in hydrogen and nitrogen doped ZnO, *Appl. Phys. Lett.*, **91**, 111903.

133. Pan, H., Luo, J., Sun, H., Feng, Y., Poh, C., and Lin, J. (2006). Hydrogen storage of ZnO and Mg doped ZnO nanowires, *Nanotechnology*, **17**, 2963–2967.

134. Young, S.J., Li, L.W., Fang, T.H., Hsueh, T.J., Meen, T.H., and Chen, I.C. (2007). Nanoscale mechanical characteristics of vertical ZnO nanowires grown on ZnO:Ga/glass templates, *Nanotechnology*, **18**, 225603.

135. Campbell, I.H., and Fauchet, P.M. (1986). The effects of microcrystal size and shapre on the one phonon Raman spectra of crystalline semiconductors. *Solid State Commun.*, **58**, 739–741.

136. Serrano, J., Manjón, F.J., Romero, A.H., Ivanov, A., Cardona, M., Lauck, R., Bosak, A., and Krisch, M. (2010). Phonon dipersion relations of zinc oxide: Inelastic neutron scattering and ab initio calculations, *Phys. Rev. B*, **81**, 174304.

137. Jie, J., Wang, G., Chen, Y., Han, X., Wang, Q., Xu, B., and Hou, J.G. (2005). Synthesis and optical properties of well-aligned ZnO nanorod array on an undoped ZnO film, *Appl. Phys. Lett.*, **86**, 031909.

138. Pan, C., Hsu, H., Cheng, H., Wu, C., and Hsieh, W. (2007). Structural and optical properties of ZnMgO nanostructures formed by Mg in-diffused ZnO nanowires, *J. Solid State Chem.*, **180**, 1188–1192.

139. Šćepanović, M., Grujić-Brojčin, M., Vojisavljević, K., Bernik, S., and Srećković, T. (2010). Raman study of structural disorder in ZnO nanopowders, *J. Raman Spectrosc.*, **41**, 914–921.

140. Wang, R.P., Xu, G., and Jin, P. (2004). Size dependence of electron–phonon coupling in ZnO nanowires, *Phys. Rev. B*, **69**, 113303.

141. Fonoberov, V.A., and Balandin, A.A. (2004). Interface and confined optical phonns in wurtzite nanocrystals, *Phys. Rev. B*, **70**, 233205.

142. Pung, S.Y., Choy, K.L., Vinogradov, E.A., Novikova, N.N., and Yakovlev, V.A. (2010). Stuctural and infrared properties of zinc oxide film and nanowires, *J. Cryst. Growth*, **312**, 2220–2225.

143. Zeng, H., Cai, W., Cao, B., Hu, J., Li, Y., and Liu, P. (2006). Surface optical phonon Raman scattering in Zn/ZnO core–shell structured nanoparticles, *Appl. Phys. Lett.*, **88**, 181905.

144. Demangeot, F., Paillard, V., Chassaing, P.M., Pagès, C., Kahn, L., Maisonnat, A., and Chaudret, B. (2006). Experimental study of LO phonons and excitons in ZnO nanoparticles produced by room-temperature organometallic synthesis, *Appl. Phys. Lett.*, **88**, 071821.

145. Park, S., Minegishi, T., Oh, D., Lee, H., Taishi, T., Park, J., Jung, M., Chang, J., Im, I., Ha, J., Hong, S., Yonenaga, I., Chikyow, T., and Yao, T. (2010). High-quality p-type ZnO films grown by co-doping of N and Te on Zn-face ZnO substrates, *Appl. Phys. Express*, **3**, 031103.

146. Tsukazaki, A., Kubota, M., Ohtomo, A. Onuma, T., Ohtani, K., Ohno, H., Chichibu, S.F., and Kawasaki, M. (2005). Blue light-emitting diode based on ZnO, *Jpn. J. Appl. Phys.*, **44**, L643–L645.

147. Tsukazaki, A., Ohtomo, A. Onuma, T., Ohtani, M., Makino, T., Sumiya, M., Ohtani, K., Chichibu, S.F., Fuke, S., Segawa, Y., Ohno, H., Koinuma, H., and Kawasaki, M. (2005). Repeated temperature modulation epitaxy for p-type doping and light-emitting diode based on ZnO, *Nat. Mater.*, **4**, 42–46.

148. Chu, Sh., Olmedo, M., Yang, Z., Kong, J., and Liu, J. (2008). Electrically pumped ultraviolet ZnO diode lasers on Si, *Appl. Phys. Lett.*, **93**, 181106.

149. Yang, Z., Chu, S., Chen, W.V., Li, L., Kong, J., Ren, J., Yu, Paul K.L., and Liu, J. (2010). ZnO:Sb/ZnO:Ga light emitting diode on c-plane sapphire by molecular beam epitaxy, *Appl. Phys. Express*, **3**, 032101.

150. Nakahara, K., Akasaka, S., Yuji, H., Tamura, K., Fujii, T., Nishimoto, Y., Takamizu, D., Sasaki, A., Tanabe, T., Takasu, H., Amaike, H., Onuma, T., Chichibu, S.F., Tsukazaki, A., Ohtomo, A., and Kawasaki, M. (2010).

Nitrogen doped $Mg_xZn_{1-x}O$/ZnO single heterostructure ultraviolet light-emitting diodes on ZnO substrates, *Appl. Phys. Lett.*, **97**, 013501.

151. Chu, Sh., Wang, G., Zhou, W., Lin, Y., Chernyak, L., Zhao, J., Kong, J., Li, L., Ren, J., and Liu, J. (2011). Electrically pumped waveguide lasing from ZnO nanowires, *Nat. Nanotechnol.*, **6**, 506–510.

152. Cao, B.Q., Lorenz, M., Brandt, M., von Wenckstern, H., Lenzner, J., Biehne, G., and Grundmann M. (2008). P-type conducting ZnO:P microwires prepared by direct carbothermal growth, *Phys. Stat. Sol. RRL*, **2**, 37–39.

153. Lorenz, M., Cao, B.Q., Zimmermann, G., Biehne, G., Czekalla, C., Frenzel, H., Brandt, M., von Wenckstern, H., and Grundmann, M. (2009). Stable p-type ZnO:P nanowire/n-type ZnO:Ga film junctions, reproducibly grown by two-step pulsed laser deposition, *J. Vac. Sci. Technol. B*, **27**, 1693–1697.

154. Dietrich, C.P., Brandt, M., Lange, M., Kupper, J., Böntgen, T., von Wenckstern, H., and Grundmann, M. (2011). Defect properties of ZnO and ZnO:P microwires, *J. Appl. Phys.*, **109**, 013712.

155. Zimmler, M.A., Voss, T., Ronning, C., Capasso, F. (2009). Exciton-related electroluminescence from ZnO nanowire light-emitting diodes, *Appl. Phys. Lett.*, **94**, 241120.

156. Zhang, J.-Y., Zhang, Q.-F., Deng, T.-S., and Wu, J.-L. (2009). Electrically driven ultraviolet lasing behavior from phosphorus-doped p-ZnO nanonail array/n-Si heterojunction, *Appl. Phys. Lett.*, **95**, 211107.

157. Richters, J.-P., Kalden J., Gnauck, M., Ronning, C., Dietrich, C.P., von Wenckstern, H., Grundmann, M., Gutowski, J., and Voss, T. (2012). Modal gain and its diameter dependence on single-ZnO micro- and nanowires, *Semicond. Sci. Technol.*, **27**, 015005.

158. Zhou, J., Gu, Y., Hu, Y., Mai, W., Yeh, P.-H., Bao, G., Sood, A.K., Polla, D.L., and Wang, Z.L. (2009). Gigantic enhancement in response and reset time of ZnO UV nanosensor by utilizing Schottky contact and surface functionalization, *Appl. Phys. Lett.*, **94**, 191103.

159. Zhai, T., Li, L., Wang, X., Fang, X., Bando, Y., and Golberg, D. (2010). Recent developments in one-dimensional inorganic nanostructures for photodetectors, *Adv. Funct. Mater.*, **20**, 4233–4248.

160. Soci, C. Zhang, A. Xiang, B., Dayeh, S.A., Aplin, D.P.R., Park, J., Bao, X.Y., Lo, X.Y., and Wang, D. (2007). ZnO nanowire UV photodetectors with high inernal gain, *Nano Lett.*, **7**, 1003–1009.

161. Lu, C.-Y., Chang, S.-J., Chang, S.-P., Lee, C.-T., Kuo C.-F., Chang H.-M., Chiou Y.-Z., Hsu C.-L., and Chen I.-C. (2006). Ultraviolet photodetectors with

ZnO nanowires prepared on ZnO:Ga/glass templates, *Appl. Phys. Lett.*, **89**, 153101.

162. Ghosh, R., Dutta, M., and Basak, D. (2007). Self-seeded growth and ultraviolet photoresponse properties of ZnO nanowire arrays, *Appl. Phys. Lett.*, **91**, 073108.

163. Law, J.B.K., and Thong, J.T.L. (2006). Simple fabrication of a ZnO nanowire photodetector with a fast photoresponse time, *Appl. Phys. Lett.*, **88**, 133114.

164. Li, Y., Della Valle, F., Simonnet, M., Yamada I., and Delaunay, J.-J. (2009). Competitive surface effects of oxygen and water on UV photoresponse of ZnO nanowires, *Appl. Phys. Lett.*, **94**, 023110.

165. Liao, Z.-M., Zhang, H.-Z., and Yu, D.-P. (2010). Ultralow-frequency photocurrent oscillation in ZnO nanowires, *Appl. Phys. Lett.*, **97**, 033113.

166. Liao, Z.-M., Xu, J., Zhang, J.-M., and Yu, D.-P. (2008). Photovoltaic effect and charge storage in single ZnO nanowires, *Appl. Phys. Lett.*, **93**, 023111.

167. Cheng, G., Li, Z., Wang, S., Gong, H., Cheng, K., Jiang, X., Zhou, S., Du, Z., Cui, T., and Zou, G. (2008). The unsaturated photocurrent controlled by two-dimensional barrier geometry of a single ZnO nanowire Schottky photodiode, *Appl. Phys. Lett.*, **93**, 123103.

168. Lin, D., Wu, H., Zhang, W., Li, H., and Pan, W. (2009). *Appl. Phys. Lett.*, **94**, 172103.

169. Das, S.N., Moon, K.-J., Kar, J.P., Choi, J.-H., Xiong, J., Lee, T.I., and Myoung, J.-M. (2010). ZnO single nanowire-based UV detectors, *Appl. Phys. Lett.*, **97**, 022103.

170. Yang, Q., Guo, X., Wang, W., Zhang, Y., Xu, S., Lien, D.H., and Wang, Z.L. (2010). Enhancing sensitivity of a single ZnO micro-/nanowire photodetector by Piezo-phototronic effect, *ACS Nano*, **4**, 6285–6291.

171. Liu, K., Sakurai, M., Liao, M., and Aono, M. (2010). Giant improvement of the performance of ZnO nanowire photodetectors by Au nanoparticles, *J. Phys. Chem. C*, **114**, 19835–19839.

172. Newton, M.C., and Shaikhaidarov, R. (2009). ZnO tetrapod p–n junction diodes, *Appl. Phys. Lett.*, **94**, 153112.

173. Wang, G., Chu, S., Zhan, N., Lin, Y., Chernyak, L., and Liu, J. (2011). ZnO homojunction photodiodes based on Sb-doped p-type nanowire array and n-type film for ultraviolet detection, *Appl. Phys. Lett.*, **98**, 041107.

174. Guo, Z., Zhao, D., Liu, Y., Shen, D., Zhang, J., and Li, B. (2008). Visible and ultraviolet light alternative photodetector based on ZnO nanowire/n-Si heterojunction, *Appl. Phys. Lett.*, **93**, 163501.

175. Um, H.-D., Moiz, S.A., Park K.-T., Jung, J.-Y., Jee, S.-W., Ahn, C.H., Kim, D.C., Cho, H.K., Kim, D.-W., and Lee, J.-H. (2011). Highly selective spectral response with enhanced responsivity of n-ZnO/p-Si radial heterojunction nanowire photodiodes, *Appl. Phys. Lett.*, **98**, 033102.

176. Zhou, H., Fang, G., Yuan, L., Wang, C., Yang, X., Huang, H., Zhou, C., and Zhao, H. (2009), Deep ultra violet and near infrared photodiode based on n-ZnO/p-silicon nanowire heterojunction fabricated at low temperature, *Appl. Phys. Lett.*, **94**, 013503.

177. Chivukula, V., Ciplys, D., Shur, M., and Dutta, P. (2010). ZnO nanoparticle surface acoustic wave UV sensor, *Appl. Phys. Lett.*, **96**, 233512.

178. Li, C., Bando, Y., Liao, M., Koide, Y., and Golberg, D. (2010). Visible-blind deep-ultraviolet Schottky photodetector with a photocurrent gain based on individual Zn_2GeO_4 nanowire, *Appl. Phys. Lett.*, **97**, 161102.

179. Yan, C., Singh, N., and Leea, P.S. (2010). Wide-bandgap Zn_2GeO_4 nanowire networks as efficient ultraviolet photodetectors with fast response and recovery time, *Appl. Phys. Lett.*, **96**, 053108.

180. Neumann, U., Grunwald, R., Griebner, U., Steinmeyer, Seeber, W. (2004). Second-harmonic efficiency of ZnO nanolayers, *Appl. Phys. Lett.*, **84**, 170–172.

181. Neumann, U., Grunwald, R., Griebner, U., Steinmeyer, G., Schmidbauer, M., Seeber, W. (2005). Second-harmonic performance of a-axis-oriented ZnO nanolayers on sapphire substrates, *Appl. Phys. Lett.*, **87**, 171108.

182. Johnson, J.C., Knutsen, K.P., Yan, H., Law, M., Zhang, Y., Yang, P., and Saykally, R.J. (2004). Ultrafast carrier dynamics in single ZnO nanowire and nanoribbon lasers, *Nano Lett.*, **4**, 197–204.

183. Prasanth, R., van Vugt, L.K., Vanmaekelbergh, D.A.M., and Gerritsen, H.C. (2006). Resonance enhancement of optical second harmonic generation in a ZnO nanowire, *Appl. Phys. Lett.*, **88**, 181501.

184. Zhang, C.F., Dong, Z.W., You, G.J., Zhu, R.Y., Qiana, S.X., Deng, H., Cheng, H., and Wang, J.C. (2006). Femtosecond pulse excited two-photon photoluminescence and second harmonic generation in ZnO nanowires, *Appl. Phys. Lett.*, **89**, 042117.

185. Zhang, C.F., Dong, Z.W., Liu, K.J., Yan, Y.L., Qian, S.X., Deng, H. (2007). Multiphoton absorption pumped ultraviolet stimulated emission from ZnO microtubes, *Appl. Phys. Lett.*, **91**, 142109.

186. Zhang, C., Zhang, F., Qian, S., Kumar, N., Hahm, J., and Xu, J. (2008). Multiphoton absorption induced amplified spontaneous emission from biocatalyst-synthesized ZnO nanorods, *Appl. Phys. Lett.*, **92**, 233116.

187. Zhang, Y., Zhou, H., Liu, S.W., Tian, Z.R., and Xiao, M. (2009). Second-harmonic whispering-gallery modes in ZnO nanotetrapod, *Nano Lett.*, **9**, 2109–2112.

188. Das, S.K., Bock, M., O'Neill, C., Grunwald, R., Lee, K.M., Lee, H.W., Lee, S., and Rotermund, F. (2008). Efficient second harmonic generation in ZnO nanorod arrays with broadband ultrashort pulses, *Appl. Phys. Lett.*, **93**, 181112.

189. Dufft, D., Rosenfeld, A., Das, S.K., Grunwald, R., and Bonse, J. (2009). Femtosecond laser-induced periodic surface structures revisited: A comparative study on ZnO, *J. Appl. Phys.*, **105**, 034908.

190. Park, J.S., Yamazaki, Y., Takahashi, Y., Hong, S.K., Chang, J.H., Fujiwara, T., and Yao, T. (2009). Origin of second-order nonlinear optical response of polarity-controlled ZnO films, *Appl. Phys. Lett.*, **94**, 231118.

191. Zheng, C.C., Xu, S.J., Ning, J.Q., Chen, Y.N., Lu, X.H., Ling, C.-C., Che, C.M., Gao, G.Y., Hao, J.H., Brauer, G., and Anwand, W. (2011). Ion-implantation induced nano distortion layer and its influence on nonlinear optical properties of ZnO single crystals, *J. Appl. Phys.*, **110**, 083102.

192. Das, S.K., Biswas, M., Byrne, D., Bock, M., McGlynn, E., Breusing, M., and Grunwald, R. (2010). Multiphoton-absorption induced ultraviolet luminescence of ZnO nanorods using low-energy femtosecond pulses, *J. Appl. Phys.*, **108**, 043107.

Chapter 3

Piezoelectric Nanogenerator Based on ZnO Nanomaterials

Jinhui Song

Department of Metallurgical and Materials Engineering, University of Alabama, Tuscaloosa, AL 35487, USA

jhsong@eng.ua.edu

This chapter introduces a nanoscale generator based on piezoelectric and semiconducting materials, mainly ZnO nanostructured materials, which include the first nanogenerator based on conductive atomic force microscopy (AFM) technique scanning on ZnOnanowire (NB) arrays, power generation process of a single-crystal ZnO wire/belt, direct current nanogenerator, high power output nanogenerator, and lateral designed nanogenerator and its building up. The detailed mechanism of energy harvesting of piezoelectric and semiconducting nanostructured materials is well discussed. The methods and designs for improving the output power are illustrated. Next-generation tine power source is highly expected from nanogenerators.

Zinc Oxide Nanostructures: Advances and Applications
Edited by Magnus Willander

ISBN 978-981-4411-33-2 (Hardcover), 978-981-4411-34-9 (eBook)
www.panstanford.com

3.1 Introduction: An Explanation of the Motivation for Developing Nanogenerator

Although researchers have fabricated and are fabricating nanoscale devices by using NW and other nanostructured materials, which offer low power consumption and ultra-small size, etc., novel properties powering those small devices are still challenging. Normally, a battery is still the power solution for nanodevices, although those tiny devices require very small amount power to work. The power source adds significant volume to the nanodevice, which will definitely lower the advantage of ultra-small size. Especially, for implantable application, such as living cell glucose sensor and muscle strain sensor, small-size power sources that can harvest energy from the environment are critically anticipated. Consequently, researchers are developing innovative technologies to harvest various forms of energy and convert them into electricity. Among our living environment, mechanical energy is the most common energy, e.g., blood flow, human body movement, machine movement, ocean tides, and winds. It is highly desired to develop nanotechnology to harvest the wasted mechanical energy from the surroundings and convert it into electric energy and finally power nanodevices, MEMS, and even personal electronics. Building working independent self-powered wireless nanosystem is the ultimate goal of nanogenerator research.

3.2 First Nanogenerator

The first nanogenerator is demonstrated by using conductive AFM scanning on aligned ZnO NW arrays [1]. The mechanism of nanogenerator relies on the piezoelectric and semiconducting dual properties of ZnO NW arrays as well as the Schottky barrier formed between ZnO NW and metal AFM tip.

The sample is ZnO NW arrays grown on *c* plane–oriented sapphire substrate by the vapor–liquid–solid (VLS) process. As shown in Fig. 3.1a. The ZnO NW grows along *c*-direction with a diameter of ~50 nm and length 200–500 nm as shown in Fig. 3.1b. The epitaxial relation between ZnO and sapphire substrate allows a thin, continuous layer of ZnO to form a common electrode at the bottom of all the NW arrays grown on the substrate. Gold catalyst that induces the NW growth vaporizes and a small amount of the NWs contain the gold residue. However, the small gold particle is so small and only covers a small fraction of the NW's top surface as indicated in the inset of Fig. 3.1b. After the first scanning of the sample, those survived particles will be scraped off. So for the power generation process, the NW has no gold particle on the top at all. The sample is connected in the conductive AFM with silver paste on the corner of conductive ZnO layer, which serves as a large electrode connecting all the NWs for transport measurement.

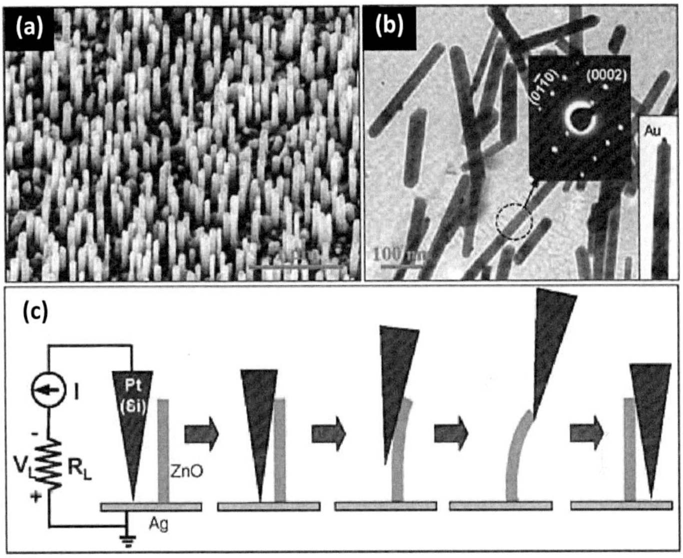

Figure 3.1 (a) SEM image of aligned ZnO NW arrays grown on sapphire substrate. (b) TEM image of the ZnO NWs and insets are diffraction pattern and enlarged image of the NW with small gold particle on the top. (c) Mechanism of the first nanogenerator by using conductive AFM scanning on ZnO NW.

The AFM tip is made of silicon and coated with a layer of platinum. The tip has an apex angle of 70° and the rectangular cantilever has a calibrated normal spring constant of 0.76 N/m. As shown in Fig. 3.1c, the AFM works in contact mode with a small constant force. As the tip scans over a NW, the height of the tip is adjusted according to the surface morphology. When AFM tip scans over the top of the NW, the deformation of the NW and the corresponding bending force that is applied by the AFM tip on the NW both attain their maximum. As the NW is released, AFM tip touch with the substrates again. During the entire scanning process, the voltage drop across a resistor ($5 \times 10^8\ \Omega$) is continuously monitored as indicated in Fig. 3.1c and there is no external voltage is applied in any stage of the experiment.

Both the topography and voltage signal images are captured as shown in Figs. 3.2a,b, respectively. From the topography, each bright triangle is a NW, which is the result of image convolution between pyramid shape AFM tip and ZnO NW. The blur is caused by the continuously bending and releasing NW. From the topography, more information can be obtained, such as the NW's length, maximum bending distance, and density. The 3D electrical signal image is composed of numerous sharp peaks as shown in Fig. 3.1b. Each electrical peak is corresponding to one ZnO NW. The peak amplitude is around 6–9 mV. Because there is no external voltage applied in the entire experiment, all the peaks are the electrical pulses generated by the ZnO NWs. This is the first experiment that converts mechanical energy (AFM tip induced mechanical deformation) into electricity (electrical pulses across the external resister) by using nanoscale material.

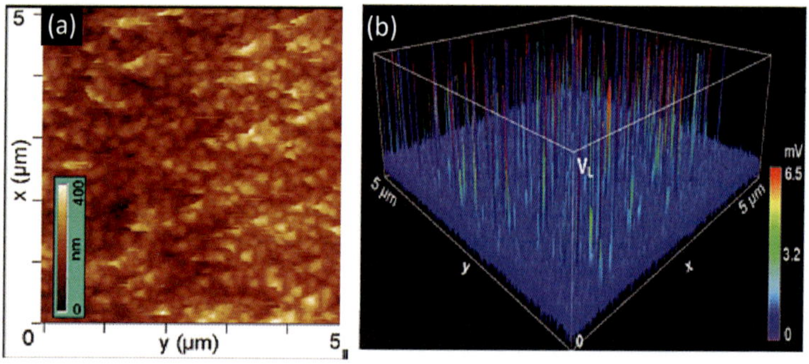

Figure 3.2 (a) Topography of conductive AFM scanned on ZnO NW arrays. (b) 3D electrical signal image corresponding to the scanning.

The physical principle for nanogenerator arises from the piezoelectric and semiconducting ZnO NW. For a vertical ZnO NW as shown in Fig. 3.3a, the deflection of the NW by AFM tip creates a strain field, which is simulated as shown in Fig. 3.3b. Negative strain is formed in the compressed side of the NW, while positive strain distributes along the stretched side of the NW. Due to the *c*-direction–orientated ZnO NW, an electrical field is created along the strain (Fig. 3.3c), which is the direct result of piezoelectric property. The electrical potential along the NW is shown in Fig.3.3d, the bottom electrode is grounded. Under the first-order approximation, the electric potential distribution from the compressed side to the stretched side of NW is V_s^- to V_s^+ across the width of the NW at the top end as shown in Fig. 3.3d. The potential is created by the relative displacement of the Zn^{2+} cations with respect to the O^{2-} anions due to piezoelectric effect in the wurtzite crystal structure.

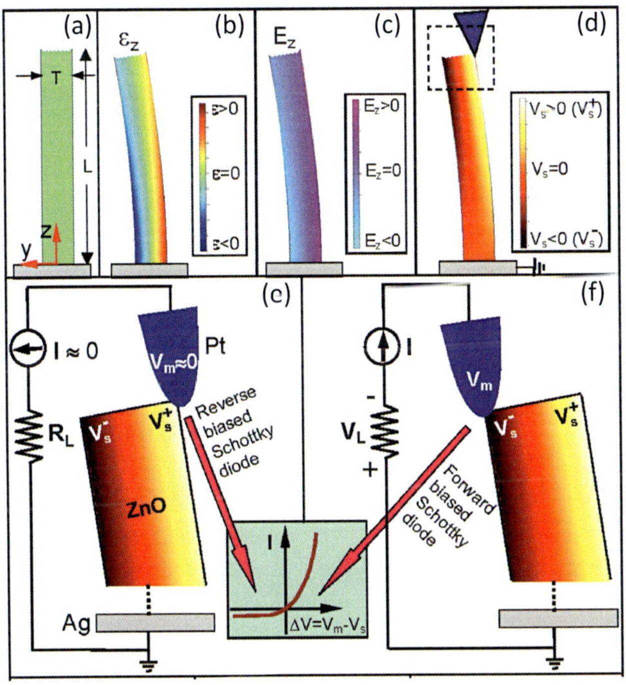

Figure 3.3 (a) Schematic definition of ZnO NW and the coordination system. (b) Strain distribution along the NW under AFM deformation. (c) The corresponding electrical field distribution. (d) the corresponding electrical potential distribution. (e, f) Metal and semiconductor contacts between Pt-coated AFM tip and the semiconducting ZnO NW.

The contacts at the top and the bottom of the NW are not symmetric; the bottom contact is the ZnO thin film in contact with silver paste, so the effective contact is between ZnO and Ag. The electron affinity of ZnO is about 4.5 eV [2] and the work function of Ag is approximately 4.2 eV, there is no barrier at the interface, so the ZnO–Ag contact is ohmic and the *I*–*V* characterization is linear. At the top of the NW, Pt has a work function of 6.1 eV, which is larger than the electron affinity of ZnO, so Schottky barrier forms and dominates the entire electrical transport process. When AFM tip that induces the deformation is in contact with stretched side surface of NW, the positive potential reversely biases the Schottky barrier formed by the Pt-coated tip and n-type ZnO NW and little current flows across the interface. In the second step, when AFM tip is in contact with the compressed side surface of the NW where the negative potential distributes, Schottky barrier is suddenly forwardly biased and it produces a sudden increase in the output electrical current, which results in an electrical pulse that can be detected by the building in voltage meter. The whole electrical energy generating process described can be schematically demonstrated by Fig. 3.3e,f.

We now estimate the possibility of powering nanodevices with the NW-based power generator. The energy output by one NW in one discharge event is ~0.05 fJ, and the output voltage on the load is ~8 mV. For an NW of typical resonance frequency ~10 MHz, the output power of the NW would be ~0.5 pW. If the density of NWs per unit area on the substrate is $20/\mu m^2$, the output power density is ~10 $pW/\mu m^2$. By choosing a NW array of size 10 μm × 10 μm, the power generated may be enough to drive a single NW/NB/nanotube (NT) based device [3–5]. If we can find a way to induce the resonance of a NW array and output the converted power generated in each cycle of the vibration, a significantly strong power source may be possible for self-powering nanodevices. Furthermore, using acoustic wave, ultrasonic wave or hydraulic pressure/force, it is possible to generate electricity using ZnO NW arrays grown on solid substrates or flexible polymer films [6]. The principle and the nanogenerator demonstrated could be the basis for exploring new self-powering nanotechnology that harvests electricity from the environment for applications in biomedical sciences (such as implantable devices), wireless sensors, portable electronics and many more.

3.3 Visualizing Power Generation Process

In the previous section, we demonstrated the first nanogenerator by using conductive AFM scanning on ZnO NW arrays. In order to further prove the mechanism we proposed in previous section, we will visually observe the power generation process under optical microscopy [7]. A longer ZnO single-crystal wire is chosen to be manipulated under optical microscope by AFM and the details of electrical signal generating process is visualized.

We use the same conductive AFM and the same setup as depicted in the first nanogenerator, but the sample is longer ZnO wire. One end of the single ZnO wire is fixed by silver paste on noconductive intrinsic silicon substrate and the other end is left free. The wire is laid down on the substrate but a small distance is kept from the substrate to eliminate the friction especial at the affixed side as shown in Fig. 3.4a.

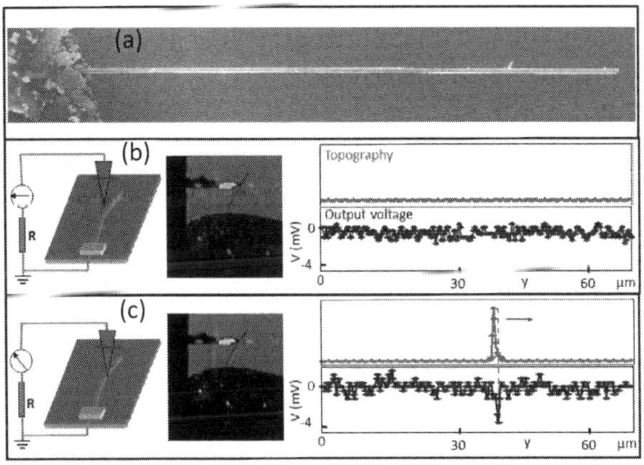

Figure 3.4 (a) SEM image of a ZnO wire with one end affixed by silver paste onto an intrinsic silicon substrate and the other end is free. (b) The AFM tip pushes the wire toward the right-hand side but does not go above and across its width, as there is no bump shown in the topography scanning profile. No output voltage is detected as shown in the blue curve of corresponding voltage output scanning profile. (c) The AFM tip pushes the wire toward the right-hand side and goes above and across its width as indicated in the appearance of a bump in the topography scanning profile. The output voltage scanning line shows a sharp negative peak. There is a delay in the output voltage peak and the bump, which is indicated by the green line.

We recorded both the topography and electrical signal images (the voltage drop across the resistor that is connected in the circuit composed of AFM tip, ZnO wire, and resistor) simultaneously. The topography image reflects the tip position change in normal direction, e.g., in perpendicular to the substrate direction. When the AFM tip scans over the ZnO wire, there is a bump in the scan line of topography. We continuously monitored the output of electrical potential across the resistor and there is no any external voltage applied during the entire experiment.

When the AFM tip scans and pushes the wire without going across the wire, which is identified by the relative smooth topography scanning line, there is no voltage output observed in the corresponding voltage signal scanning line as shown in Fig. 3.4b. As the AFM tip scans above and goes across the ZnO wire as indicated as a bump shown in the topography scanning line, a negative voltage pulse appears according to the runs over accident as shown in Fig. 3.4c.

When the AFM tip deforms the ZnO wire, the AFM tip only touches the stretched side and the positive potential reversely biases the Schottky barrier formed by the Pt-coated tip and voltage accumulates. When a bump shows in the topography scanning line, which means the AFM tip is lifted and scans across the wire and touches the compressed side, positive potential at the compressed side of the ZnO wire forwardly biases the Schottky between the tip and ZnO, an electrical pulse is suddenly generated. The overall energy transformation process of mechanical energy into electrical energy is visualized under optical microscope and further confirms the mechanism proposed in Section 3.1.

3.4 Piezoelectric Potential in NW

The detailed calculation and simulation have been carried out and show that the NW's piezoelectric potential does not depend on the z-coordinate along the NW except when it closes to the two ends. An analytical calculation on the potential distribution in ZnO NW has been carried out recently [8] besides amount of numerical simulations.

Starting from governing equations that include mechanical equilibrium equation (3.1), constitutive equation (3.2), and geometrical compatibility equation (3.3), combining with Gauss equation of electric field (3.4), four independent equations can be set up for solving the electrical potential distribution analytically.

$$\nabla \cdot \sigma = \overrightarrow{f_e^{(b)}} = 0, \tag{3.1}$$

where σ is the stress tensor.

$$\begin{cases} \sigma_p = c_{pq} - e_{kp}E_k \\ D_i = e_{iq}\varepsilon_q + k_{ik}E_k \end{cases}, \tag{3.2}$$

where c_{pq} is the linear elastic constant, e_{kp} is the linear piezoelectric coefficient, and k_{ki} is the dielectric constant.

$$e_{ilm}e_{jpq}\frac{\partial^2 \varepsilon_{mp}}{\partial x_l \partial x_q} = 0 \tag{3.3}$$

Here e_{ilm} and e_{ilm} are Levi-Civita antisymmetric tensors. For simplication, the NW bending is assumed to be small.

Suppose there are no free charges in the NW:

$$\nabla \cdot \vec{D} = \rho = 0 \tag{3.4}$$

Equations (3.1–3.4) combined with boundary conditions are a complete description of a static piezoelectric system, such as NW. the solution is too complicated to be deduced. However, a perturbation expansion of the linear equations and the results can be simply approximately deduced. The maximum potential at the surface (NW radius is $r = a$) at the tensile side and compressed side, respectively, is

$$\varphi_{\max}^{(T,C)} = \pm \frac{3}{4(k_0 + k_\perp)}\left[e_{33} - 2(1+v)e_{15} - 2ve_{31}\right]\frac{a^3}{l^3}v_{\max} \tag{3.5}$$

From Eq. (3.5), the electrostatic potential is directly related to the aspect ratio of the NW. For a ZnONW with diameter of 50 nm and length of 600 nm at a lateral bending force of 80 nN the analytical calculation result shows maximum potential is around 0.3 V. The comparison of the analytical calculation with the numerical simulation is shown in Fig. 3.5.

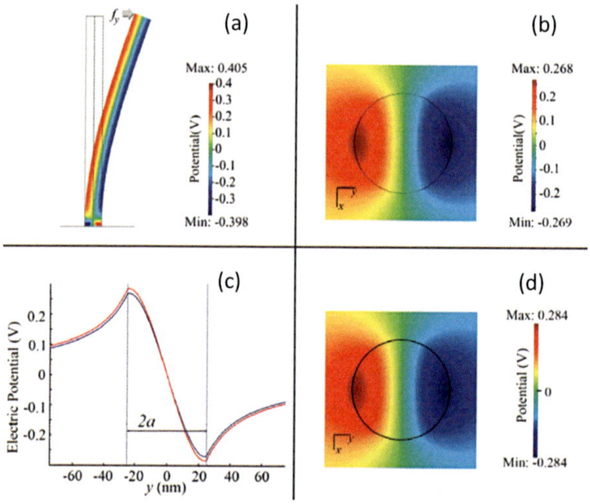

Figure 3.5 Potential distribution in a ZnO NW with diameter of 50 nm and length of 600 nm under a lateral bending force of 80 nN. (a) and (b) are side and top cross-sectional output of the piezoelectric potential in the NW given by finite element calculation using full coupled equations 3.1–3.4. (c) is the cross-sectional output of the piezoelectric potential given by analytical calculation. (d) gives a comparison of the line scan profiles from both (b) and (c).

To compare the identity of the numerical simulation with the analytical calculation, a line scan is made across the output of the cross-sectional potential along the symmetry line following the lateral deflection direction, and the results are shown in Fig. 3.5d. The difference is smaller than 6%, proving the accuracy of the analytical solution.

The elastic modulus and piezoelectric coefficient used for the calculations are adopted from the values obtained from bulk and film. Previous experimental measurements show the elastic modulus and piezoelectric coefficient of ZnO NW are much larger than bulk counterparts. If these values are used in the calculation, the potential of the NW surface would be increased by a factor of 3–4.

3.5 Direct Current Nanogenerator

The nanogenerators mentioned in previous sections are all based on conductive AFM technique. For real applications, using AFM as the mechanical energy input source must be eliminated and the power generation can be achieved in a large scale of NWs. All of the NWs are required to generate electricity simultaneously and all the electricity can be collected and output effectively. The mechanical deformation needs to be provided in a form of wave from the environment, so the nanogenerator can work independently and wirelessly. The direct current (DC) nanogenerator demonstrated here shows an innovative approach that uses ultrasonic wave to agitate and input mechanical deformation on ZnO NW arrays [9]. The design has the potential to satisfy all the requirements outlined above. This approach could be the foundation for nanogenerator to have practical application in nanotechnology.

The mechanism design of DC nanogenerator is shown in Fig. 3.6a, in which aligned ZnO NW arrays grown on sapphire or GaN substrate are covered with a zigzag shape top electrode. Top electrode is obtained by etching silicon and coating with a layer of Pt that serves as conductive electrode and forming Schottky contact at the interface with ZnO. The distance between top electrode and ZnO NW arrays is carefully adjust to keep a small distance, which is fixed and packaged by flexible epoxy. A cross-sectional SEM image of the packaged device is shown in Fig. 3.6b, displaying a "lip–teeth" relationship between the electrode and NWs. Ultrasonic wave has a frequency of 41 kHz and the mechanical deformation of the NWs is expected from the agitating ultrasonic wave. Electricity

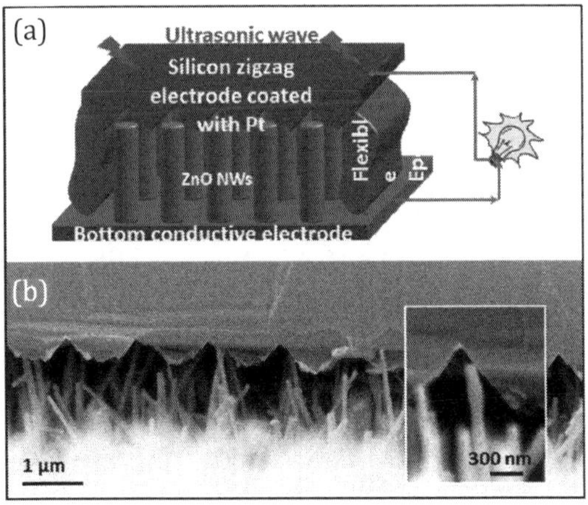

Figure 3.6 (a) Mechanism design of DC nanogenerator. (b) SEM image of the cross-section of as-fabricated DC nanogenerator.

harvested by the NWs is expected to be output simultaneously and continuously.

The working mechanism is based on the first nanogenerator in Section 3.2. The asymmetric piezoelectric-potential across the width of a ZnO NW and the Schottky barrier formed between the metal electrode and NW are the two key reasons for creating, separating, accumulating, and outputting the charges. The top zigzag electrode is to act as an array of AFM tips as indicated in Fig. 3.7a. We take NW I in Fig. 3.7 as an example to demonstrate the process of outputting electric energy generated in each cycle of the NW deflection. When the top electrode/NWs are subjected to the excitation of ultrasonic wave, the zigzag top electrode will move up and down to push the NW, which results in the deformation of NW I, creating a strain field across NW width, with inner surface of the NW being in compressive strain and outer surface being in tensile strain, respectively. Because of piezoelectric property of the NW, a piezoelectric potential forms from negative maximum value to positive maximum value across the top of the NW surface. When the trench of the top zigzig top electrode is in contact with the stretched side and bending the NW, positive potential on the stretched side of the NW reversely biases the Schottky barrier between Pt-coated top electrode and ZnO NW, resulting in the piezoelectric potential accumulation. As the further pushing of the electrode, the bent NW I will reach the other side of the adjacent tooth of the zigzag electrode as shown in Fig. 3.7b. In such a case, the electrode also contacts the compressed side of the NW, where Schottky barrier is now forwardly biased, resulting in a sudden increase in the output electric current flowing from the top electrode to the NW. This is the major power generation process. Besides, there are several possible power generation processes in this device. Figure 3.7b shows NW III is working under resonance mode agitated by the ultrasonic wave, in which the NW's compressed surface always touches the electrode and thus the Schottky barrier is always forwardly biased. NW IV works under the compression of top electrode mode. The potential at top surface of the NW is always negative so that always forwardly biases the Schottky barrier between electrode and NW, leading to the contribution of electrical current output of the nanogenerator.

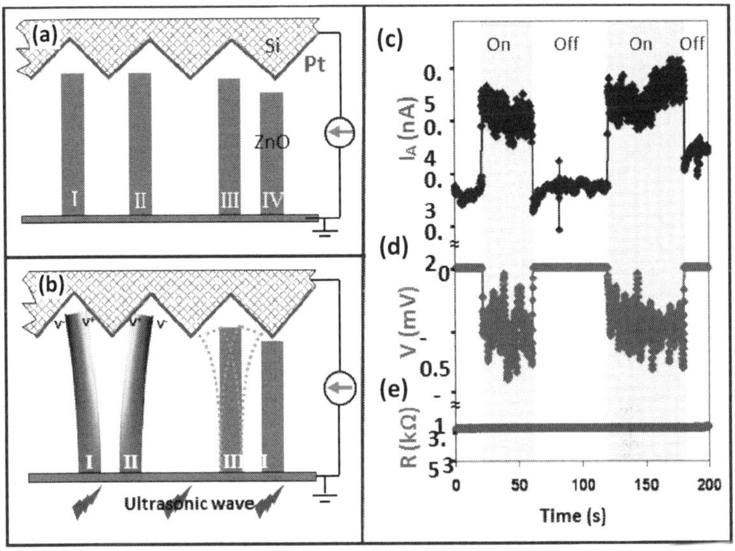

Figure 3.7 (a) Schematic illustration of the zigzag electrode and the types of the representing configurations of the NWs. (b) Working mechanism of different types of the NWs. (c) Current output of DC nanogenerator. (d) Voltage output of DC nanogenerator. (e) Resistance *R* measured during the testing process.

The current and voltage outputs of the nanogenerator are shown in Fig. 3.7c,d, respectively, with the ultra-sonic wave being turned on and off periodically. A jump of about 0.15 nA is apparent when the ultrasonic wave is turned on (yellow shaded area in Fig. 3.7c–e), and the current immediately falls back to the base line once the ultrasonic wave is turned off. Similarly, the voltage output is measured to have the same on and off tendency but with negative output of about 0.5 mV.

The resistance *R* of the entire device is also monitored with and without turning on the ultrasonic wave, as shown in Fig. 3.7e. The resistance maintains a very stable value regardless the ultrasonic wave is on or off. The evidence indicates that the change in current and voltage is not due to the variation of resistance, confirming that the current signal is truly generated by the nanogenerator.

The design shown in Fig. 3.6a has been tested in control experiments by using different materials or configurations. By replacing ZnO NWs with arrays of carbon NTs, no output current is observed as shown in Fig. 3.8a. This is because the carbon NT is not piezoelectric. Using ZnO NWs but replacing top electrode with a flat thin Pt film as shown in Fig. 3.8b, no current output can be detected either (Fig. 3.8b). This is because the design is not followed the nanogenerator mechanism. Output can be only measured when the top electrode is a zigzag shape with the use of piezoelectric NWs as shown in Fig. 3.8c. These control experiments rule out the possible contribution from electronic noise of the system and the measurement error or artifacts in generating the output current, and they do consistently support the mechanism proposed here.

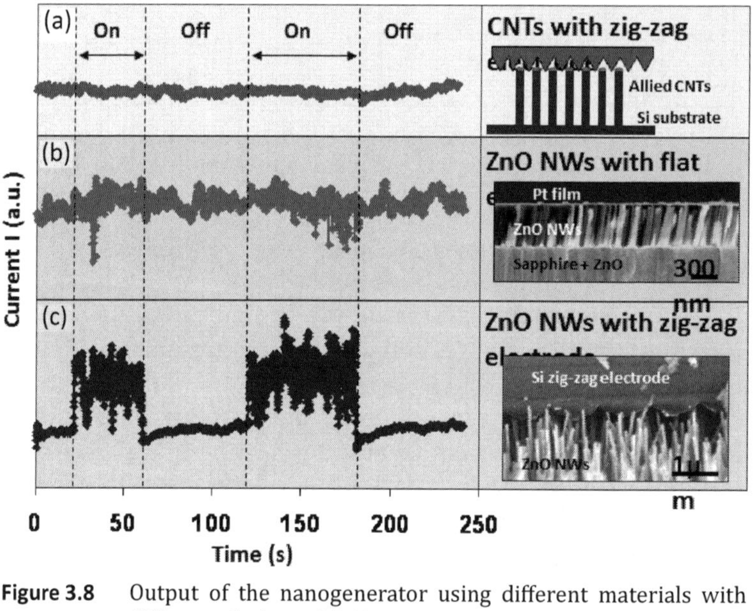

Figure 3.8 Output of the nanogenerator using different materials with different designs. (a) Nanogenerator composed with carbon NTs with a zigzag top electrode, showing no current output regardless the ultrasonic wave is on or off. (b) Nanogenerator based on ZnO NW arrays but with flat top electrode, showing no current output when the ultrasonic-wave is turned on. (c) Nanogenerator composed with piezoelectric ZnO NW arrays and zigzag top electrode, showing detectable current output when the ultrasonic wave is turned on.

In comparison to the AFM-based first nanogenerator, the DC nanogenerator has achieved three major objectives: replacing AFM tip with the ultrasonic wave for inducing elastic deformation; integrating an array of tips into a zigzag electrode for simultaneous creating and outputting electricity generated by many NWs; and the obtaining fairly stable and continuous DC output. As limited by the smaller degree of the ultrasonic wave induced-elastic deformation to the NWs in comparison with that induced by the AFM tip, the output voltage is naturally smaller. Thus, high-power-output nanogenerator is highly desired.

3.6 High-Power-Output Nanogenerator

In this section, a design of DC nanogenerators connected in serial will be discussed to improve the power output [10]. The nanogenerator is composed of hexagonal-prism-shape ZnO NW arrays and paired nanobrushes made of pyramid-shape metal-coated ZnO nanotip (NTP). Both of the ZnO nanostructures are synthesized by a wet chemical method at a temperature lower than 100°C on the two surfaces of a common silicon substrate. One piece of such structure is put in close with another to form a layer-by-layer matched brush architecture, then DC current is generated by introducing ultrasonic wave. A four-layer integrated NG can output power at a density of 0.11 μW/cm^2 at a voltage of 62 mV.

As shown in Fig. 3.9, the Si substrate is coated with 100 nm Al_2O_3 film on both sides to form insulating layers to ensure independent operations of the nanogenerators to be built in the adjacent layers. A 20 nm Cr and 50 nm ZnO layers are coated on both sides in sequence. Cr works as both a bonding material to fix the Al_2O_3 film and a common electrode for connecting all the NWs/NTPs.

The growth of NWs is via low growth temperature and long growth time and NTPs are gown by high growth temperature and short growth time, all of which use solution method for growth. The grown results are shown in Fig. 3.10a,b, respectively, for hexagonal-prism-shaped ZnO NW arrays and pyramid-shaped ZnO NTP arrays.

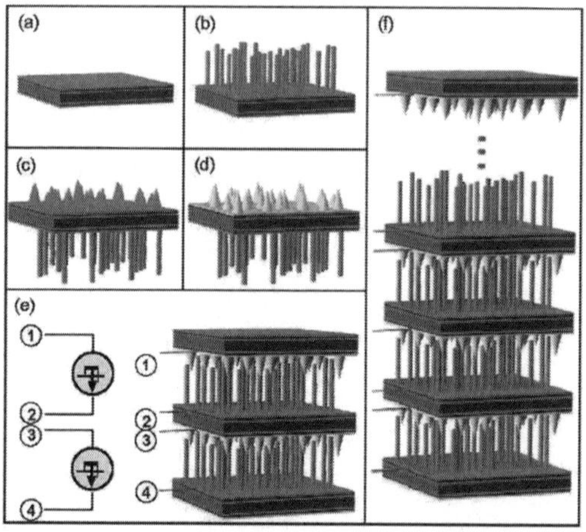

Figure 3.9 Configuration and fabrication procedures of the multilayered NGs. (a) Si substrate with double sides coated with Al_2O_3, Cr, and ZnO seed layers. (b) Grown ZnO NWs by wet chemical method on one side. (c) Growth of pyramid-shaped NTPs on the other side. (d) Coated Au conductive layer on NTPs side to form conductive tip arrays. (e) Two-layer NGs by Au-coated NTP arrays facing and interpenetrating bare ZnO NW arrays. The right-hand side is the symbol of NG. (f) Multiple-layer NG by stacking multiple layers of the wafer structures.

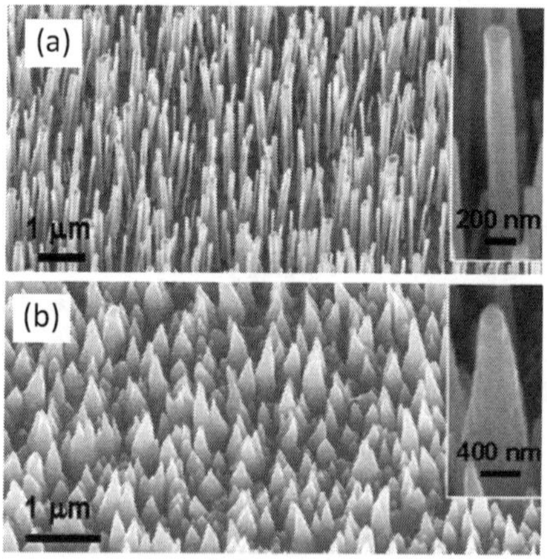

Figure 3.10 (a) SEM image of the hexagonal-prism-shaped ZnO NW arrays grown by wet chemical method. (b) SEM image of the pyramid-shaped ZnO NTP arrays grown by wet chemical method.

One NG is made by putting together two pieces of the grown wafer structures, with Au-coated NTPs partially interpenetrating into the NWs as shown in Fig. 3.9f. By continuously adding more wafer structures, multilayer NG is fabricated. Each layer is insulated from the Al_2O_3 thin film. After making the output wires, the whole unit is sealed by epoxy resin to prevent the penetration of liquid.

As indicated in Fig. 3.9e,f, the Au-coated NTPs work as the AFM tips and the ultrasonic wave is the source for inducing mechanical deformation on NWs.

To test the feasibility of integration of multilayer NGs to achieve high output voltage, a four-layer integrated NG has been fabricated and measured and the results are shown in Fig. 3.11.

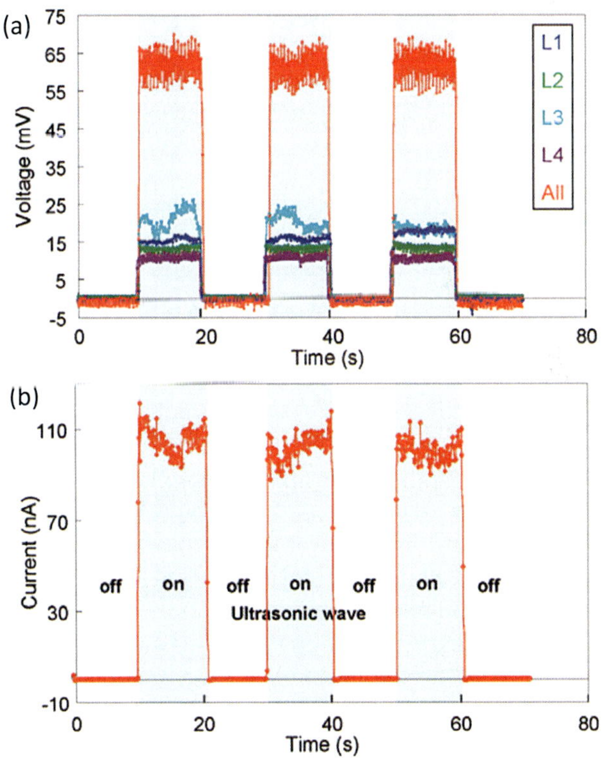

Figure 3.11 Open-circuit voltage and short-circuit current output of serially connected four-layer integrated NGs. (a) Open-circuit voltage output from individual layer and the serially connected four-layer integrated NG. (b) Short-circuit current of serially connected four-layer integrated NG.

Figure 3.11a shows the four individual layers of NGs with 11, 14, 16, and 20 mV open-circuit output voltage, respectively. The total voltage output is about 62 mV for serially connected those 4 NGs, which is about the added up of the output of the individual NG. The total current output is shown in Fig. 3.11b and is about 105 nA. The maximum power output of the four-layer integrated NG is 6.5 nW for a surface area of 6 mm^2. A power density of 0.11 $\mu W/cm^2$ is achieved.

This design presents an innovative approach for fabricating 3D integrated multilayer NGs for raising the current, voltage, and power.

3.7 Laterally Packed AC Nanogenerator

Compared with previous nanogenerators, laterally packed nanogenerator here in this section has advantages of high stability, stronger mechanical robustness, long lifetime and improved environmental adaptability. The laterally packed nanogenerator works under low mechanical deformation rate and can produce alternating current (AC) or voltage output [11].

The fabrication process of laterally packed AC nanogenerator includes a piezoelectric fine wire (PFW) lies on a flexible substrate and is fixed to electrode at both ends as shown in Fig. 3.12a and the whole unit then is packed by flexible epoxy for protection as Fig. 3.12c shows. When the substrate is bent, a tensile strain of 0.05–0.1% will be induced in the wire as schematically shown in Fig. 3.12b, leading to a piezoelectric voltage drop across the wire and forcing electrons to run through outer circuit. When the substrate is released, electrons will flow back without the piezoelectric voltage potential drop. An alternating current is produced when periodically bend and release the wire.

The output is characterized by measuring short-circuit current and open-circuit voltage of the lateral wire generator (LWG). When the current meter is forwardly connected to a LWG, a positive voltage/current pulse is detected during fast stretching of the flexible substrate as shown in Fig. 3.13a. A corresponding negative pulse for fast releasing the substrate is also detected. The output voltage of 20–50 mV and the output current of 400–750 pA can be generated by the LWG made of ZnO wire with diameter of 4 μm and length of 200 μm. The heights of the current peaks for the stretch and release show different, possibly because of various straining rates. However, the area under the voltage or current profile is the same, which means the total charges transported in the two processes remain the same.

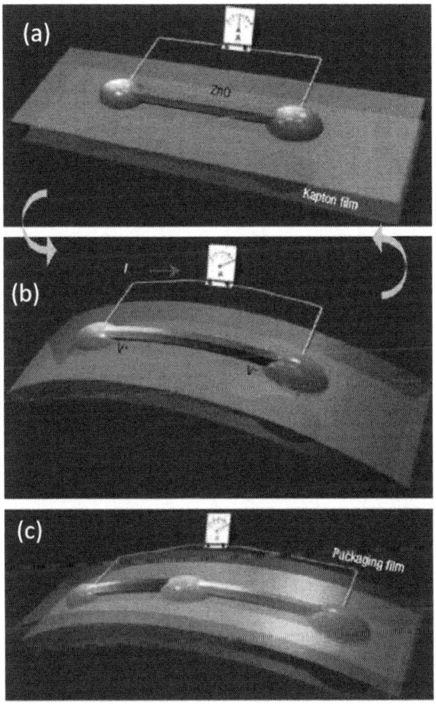

Figure 3.12 Design of a piezoelectric fine wire (PFW) generator on a flexible substrate. (a) A PFW fixed by electrode on Kapton flexible substrate. (b) Mechanical bending the flexible substrate and inducing tensile strain and corresponding piezoelectric potential across the PFW, producing current in the circuit. (c) Packaging serially connected PFWs to form AC nanogenerator.

When the LWG is reversely connected to the measurement system, the voltage and current pulses are also reversed as shown in Fig. 3.13b. The nonsymmetric output of the LWG before and after switching the connection is likely caused by a constant bias current in the measurement system, which is added up in the forward connection and subtracted from the generated current in the reverse connection.

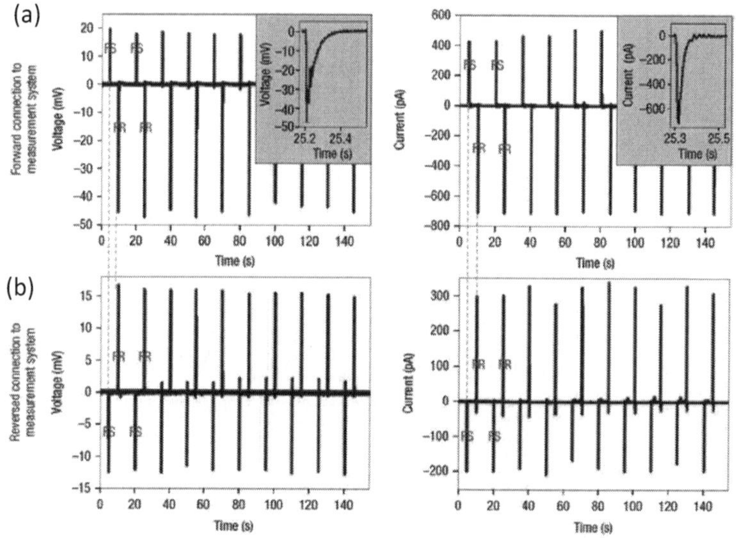

Figure 3.13 Electrical output of a lateral wire generator (LWG). (a) Output of LWG forwardly connected to the measurement system. (b) Output of LWG reversely connected to the measurement system.

The LWG is a flexible power generator that has the advantage of stability, robustness, life time in comparison with nanogenerator made from aligned NW arrays. The output voltage is about 15–100 times higher than that of DC nanogenerator.

3.8 High-Output Lateral Nanogenerator

In this section, a flexible high output nanogenerator (HONG) [12], fabricated by scalable sweeping-printing-method, will be introduced, possessing open-circuit voltage of up to 2.03 V and a peak output power density of 11 mV/cm^3. The generated electricity is stored in capacitors via bridge circuit and then lights up commercial light-emitting diode (LED), which demonstrates the progress of using nanogenerator to building self-powered devices by harvesting energy from the environment.

Based on the power generation technique of ZnO fine wire bounded on flexible substrate as described in Section 3.7, hundreds of thousands of horizontally aligned NWs have been scaled up to form integrated nanogenerator arrays.

The fabrication process includes two steps. First, the aligned nanowire arrays grown on Si substrate by vapor deposition method are transferred onto a receiving flexible substrate to form horizontally aligned arrays. Second, parallel electrode strips are deposited on the flexible substrate perpendicular to the aligned NWs by the lithography method.

The as-synthesized ZnO NWs on Si substrate have a length of 50 μm and diameter of 200 nm, and the growth direction along the *c*-axis as shown in Fig. 3.14b. The same growth direction of the NWs leads to the superposition of piezoelectric potential of all NWs created by bending the flexible substrate. The equipment for NW transferring is shown in Fig. 3.14a. ZnO NW arrays on Si substrate are attached on the stage 1 whose height can be accurately

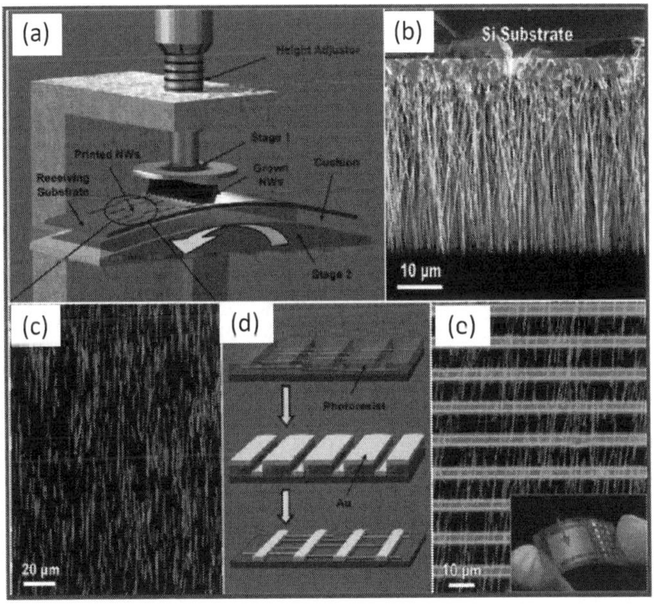

Figure 3.14 Fabrication process and structure characterization of HONG. (a) Equipment and mechanism for vertical aligned NWs transferring to laterally aligned NW arrays on flexible substrate. (b) SEM image of aligned ZnO NW arrays grown on Si wafer by vapor deposition method. (c) SEM image of the as-transferred horizontal NWs on a flexible substrate. (d) Process for fabricating electrodes on the horizontally aligned NWs. (e) SEM image of the NW arrays bounded by Au electrode. Inset is an photo of as-fabricated HONG.

adjusted. The flexible substrate for receiving NWs is attached on stage 2 by a PDMS film cushion layer to enhance the alignment of the NWs. Stage 2 can move freely in circular motion as indicated in Fig. 3.14a. The distance between the NWs and receiving substrate is precisely controlled to form a loose contact, which is the key for the successfully transferring. The as-transferred NW arrays are shown in Fig. 3.14c with an estimated density of 1.1×10^6 cm^{-2}. The lengths of the NWs indicate that not all the NWs transferred are broken off at the roots.

As shown in Fig. 3.14d, the gold electrodes that are parallel to each other are deposited perpendicularly on the aligned NWs via lithography process. The striped shape Au electrodes have a thickness of 300 nm and totally 600 rows with 10 μm in spacing as shown in Fig. 3.14e. The inset shows the effective working area of 1 cm^2, which has 3.0×10^5 NWs. A PDMS packaging on the entire structure can prevent fluid and improve the robustness.

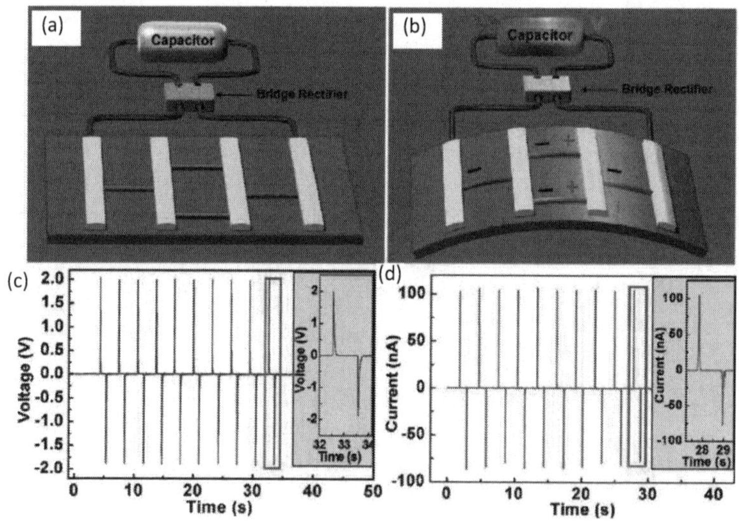

Figure 3.15 (a) Schematic structure of HONG, the output is connected with capacitor via bridge rectifier. (b) Schematic working principle of HONG. (c) Open circuit measurement of voltage output of HONG. (d) Short-circuit measurement of current output of HONG.

The working mechanism of the HONG can be illustrated by the schematic diagrams in Fig. 3.15a,b.

NWs connected in parallel contribute to the current output while NWs connected in serial superpose to the voltage output. As the flexible substrate is bent, all the NWs suffer mechanical deformation. The same growth direction of all NWs and the sweeping printing method guarantee the crystallographic orientations of the horizontal NWs are aligned along the sweeping direction. As a result, the polarity of the induced piezoelectric potential is also aligned, leading to the superposition of all the piezoelectric potential across the NWs.

Linear motor is used to bend and release the flexible HONG at a rate of 0.33 Hz. The open-circuit voltage output is shown in Fig. 3.15c and short-circuit voltage output is shown in Fig. 3.15b, respectively. At each mechanical deformation, a voltage pulse of 2.03 V and a current pulse with amplitude of 107 nA can be detected. Assuming all the NWs in HONG contribute to the output and the current produce by each NW is around 200 pA; voltage of each row of the NWs is around 3.3 mV, a peak output power density of 0.22 $\mu W/cm^2$ could be achieved, which is 20 times of that from PZT base cantilever energy harvester.

In order to demonstrate the energy storing and device driving of HONG, a switchable bridge circuit and capacitor arrays are utilized the electricity generated by HONG to power commercial LED. The inset of Fig. 3.16a shows the diagram of the circuit. HONG is operated by the linear motor at a rate of 3 Hz. Figure 3.16a shows the rectified voltage pulses generated by HONG. Rectified electrical voltage pulses continuously charge 10 capacitors connected in parallel and the voltage across the capacitors finally reached to 0.37 V. After finishing charging, capacitors are changed connection from parallel to serial and 3.7 V total voltage is used to power the commercial LED (Fig. 3.16B). The discharge process is trigged and LED is lighted up as shown in Fig. 3.16c,d.

The novel HONG technique discussed in this section demonstrates the integration of nanogenerators has the ability to enable self-powering technology and bring it into people's daily life.

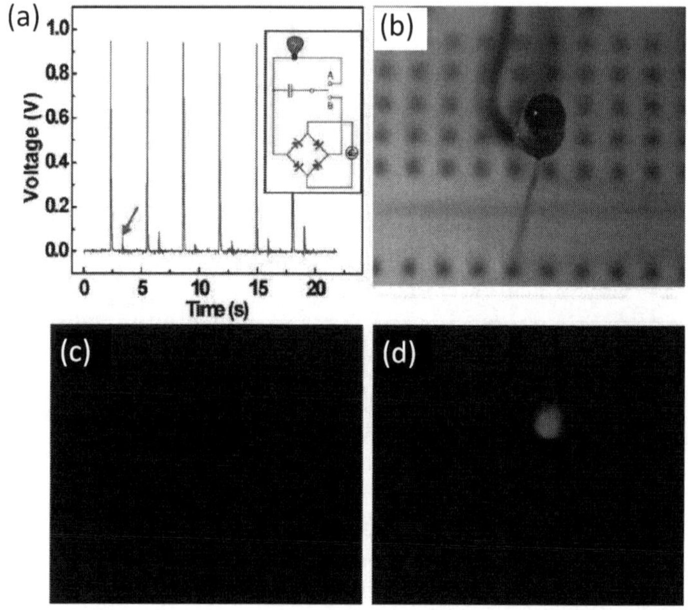

Figure 3.16 (a) Rectified electrical pulses generated by HONG. Inset is the circuit diagram for the power rectification, storage, and releasing. (b) Photo of the commercial LED connected in the HONG powered circuit. (c) Image of the LED in dim background before lighting up. (d) Image of the LED in dim background at the moment when it is lit up by the electricity generated from the HONG.

3.9 Summary

This chapter describes the nanogenerator's principle and different configuration, designs. Starting with first conductive AFM-based nanogenerator based on aligned piezoelectric ZnO NW arrays, the piezoelectric and semiconducting dual properties of ZnO NW are the keys for power generation. Detailed power-generating process is visualized under optical microscopy with a special sample design. DC nanogenerator demonstrates the first prototype independent device that can harvest mechanical energy from the environment and convert it into electricity. Stacking multilayer DC nanogenerator to improve the voltage output shows the superposition works on DC nanogenerators. Lateral AC nanogenerator shows another design to harvest low-frequency mechanical energy. The lateral scale-up of nanogenerators eventually light up a commercial LED. There are other research works that have not been included in this chapter, such as a nanogenerator based on p-type ZnO NW arrays, p–n junction gated nanogenerator, and nanogenerator based on cone shape ZnO NWs.

The research on nanogenerator will eventually provide a renewable and green energy source for people to maintain sustainable development of the human civilization. The tiny size of the nanogenerator is an ideal solution for building self-powered nanosystems that can find gigantic application in the near future.

References

1. Wang, Z. L., and Song, J. H. (2006). Piezoelectric nanogenerators based on zinc oxide nanowire arrays, *Science*, **312**, 242–246.
2. Hasegawa, S., Nishida, S., Yamashita T., and Asahi H. (2005). Field electron emission from polycrystalline GaN Nanorods, *J. Ceramic Process. Res.*, **6**, 245–249.
3. Huang, Y., Duan, X. F., Cui, Y., Lauhon, L. J., Kim, K. H., and Lieber, C. M. (2001). Logic gates and computation from assembled nanowire building blocks, *Science*, **294**, 1313–1317.
4. Bachtold, A., Hadley, P., Nakanishi, T., and Dekker, C. (2001). Logic circuits with carbon nanotube transistors, *Science*, **294**, 1317–1320.
5. Chen, J., Perebeinos, V., Freitag, M., Tsang, J., Fu, Q., Liu, J., Avouris, P. (2005). Bright infrared emission from electrically induced excitons in carbon nanotubes, *Science*, **310**, 1171–1174.
6. US patent pending.
7. Song, J. H., Zhou, J., and Wang, Z. L. (2006). Piezoelectric and semiconducting coupled power generating process of a single ZnO belt/wire, *Nano Lett.*, **6**, 1656–1662.
8. Gao, Y. F., and Wang, Z. L. (2007). Electrostatic potential in a bent piezoelectric nanowire, the fundamental theory of nanogenerator and nanopiezotronics, *Nano Lett.*, **7**, 2499–250.
9. Wang, X. D., Song, J. H., Liu, J., and Wang, Z. L. (2007). Direct-current nanogenerator driven by ultrasonic waves, *Science*, **316**, 102–105.
10. Xu, S., Wei, Y. G., Liu, J., Yang, R. S., and Wang, Z. L. (2008). Integrated multilayer nanogenerator fabricated using paired nanotip-to-nanowire brushes, *Nano Lett.*, **8**, 4027–4032.

11. Yang, R. S., Qin, Y., Dai, L. M., and Wang, Z. L. (2009). Power generation with laterally-packaged piezoelectric fine wires, *Nat. Nanotechnol.*, **4**, 34–39.

12. Zhu, G., Yang, R. S., Wang, S. H., and Wang, Z. L. (2010). Flexible high-output nanogenerator based on lateral ZnO nanowire array, *Nano Lett.*, **10**, 3151–3155.

Chapter 4

Nanobiology and Nanomedical Devices Using Zinc Oxide Nanostructures

Magnus Willander and Omer Nur

Department of Science and Technology, Campus Norrköping, Linköping University, SE-601 74 Norrköping, Sweden

magwi@itn.liu.se

In general, nanostructures with relatively small sizes constitute excellent signal transduction elements when dealing with biological analytes or chemical species. The reason for this is the similarity of the order of size between nanostructures and biological analytes and chemical species. The consequence of this issue is the fact that nanostructures will possess relatively very high sensitivity for detection of the biological analytes and chemical species compared to conventional large bulk sensors. In addition, sensors based on nanostructures can operate efficiently when the detection is for situations where the available sample volume is small and the concentration of the element of analyte in question is low. This is in fact a very important advantage of nanostructure-based sensors. Among the various known nanostructures, nanowires (NWs) and nanorods (NRs), with their relatively high-surface-area-to-volume ratio, constitute excellent signal transduction elements and have other advantages due to their geometrical features. On the other hand,

Zinc Oxide Nanostructures: Advances and Applications
Edited by Magnus Willander

ISBN 978-981-4411-33-2 (Hardcover), 978-981-4411-34-9 (eBook)
www.panstanford.com

zinc oxide (ZnO), being biosafe and biocompatible, is an attractive material for implementation as an active sensor element. Moreover, among the different sensor procedures, the potentiometric-based procedure is one of the most attractive procedures for bio-environments due to the fact that it is based on measuring charge accumulation with no current passing through the media.

4.1 Introduction

Zinc oxide (ZnO) is a direct-wide-bandgap semiconductor belonging to group II–VI and has received attention in the past few years for its potential for many applications in the area of photonics, piezoelectricity, and sensing [1]. Zinc oxide possesses excellent properties for optical application due to its relatively high exciton binding energy (60 meV), its wide direct bandgap (3.34 eV), and the many deep levels that emit covering the whole visible range. These excellent optical properties imply a potential for laser and light emitting diodes [2]. Zinc oxide has also been of interest for field emission devices [3]. In addition, ZnO is characterized by high electromechanical coupling, which has been utilized to demonstrate generation of electrical energy from mechanical movements [4]. Also, ZnO is a biosafe and biocompatible material [5]. In its unintentional doping, zinc oxide grows as an n-type material with many shallow donor levels [6]. Although ZnO is known to researchers since the 1930s [7], the interest in ZnO has been fluctuating despite many excellent properties of this material. The main reason for this was mainly the lack of stable reproducible p-type doping that can be reproduced in different laboratories. Researchers have attempted to utilize the excellent properties of ZnO by combing its thin films in heterojunctions with other p-type. Nevertheless, due to lattice mismatch, high-performance device quality heterojunctions were not demonstrated. For the past few years, the interest in ZnO has intensified in many laboratories worldwide performing research on ZnO growth and devices. This intensified research concerns ZnO nanostructures mainly for many reasons. Among these reasons are the small footprint of nanostructures implying no need for lattice mismatch and also due to the self-organized growth property of ZnO, which implies that ZnO nanostructures can be grown on any substrate that is amorphous or crystalline in nature [2]. In addition, the possibility of low-temperature chemical growth of these nanostructures adds to the advantages of this material due to the low cost, possibility of large-scale production, and growth on soft substrates, e.g., sub-micrometer glass pipettes (see below).

In addition, among all one-dimensional (1D) nanostructures, ZnO has unique properties. For example, the quality of being a semiconductor and piezoelectric holds potential for the development of electromechanically coupled sensors and transducers. In fact, the possibility of electromechanical coupling provided by ZnO is of potential for wireless devices and systems. As wireless devices may allow in situ, real-time biomedical monitoring and detection, they still require a power source. Ideally, such devices should be self-powered and not dependent on a battery [4]. The human body provides numerous potential power sources—mechanical energy (such as body movement, muscle stretching, blood vessel contraction), vibrational energy (acoustic waves), chemical energy (glucose), and hydraulic energy (body fluid and blood flow)—but the challenge is their efficient conversion into electrical energy [4]. If accomplished on the nanoscale, such power sources could greatly reduce the size of integrated nanosystems for optoelectronics, chemical and biosensors, resonators, and even more systems can benefit from this [4]. Hence, ZnO-based sensors can be good candidates for developing self-powered wireless sensors for physiological environments.

Nanostructures in general are of interest for sensing applications. This is due to the similarity in size between these nanostructures and biological analytes and other chemical species. In addition, nanostructures possess relatively large surface-area-to-volume ratio. This makes nanostructures sensitive to surface modifications and hence equips them with relatively high sensitivity when used as sensors. Due to their small size and the above-mentioned advantages, nanostructure-based sensors can operate efficiently and detect low concentrations in small volumes. Such possibility is sometimes necessary, e.g., at crime locations, and it is important to mention that conventional classical sensors based on large-area electrodes cannot perform in such way. This is because for classical large-area sensors, the volume needed for detection is inversely proportional to the concentration of the item in question [8]. Generally, electro- and biosensors are devices in which electrochemical recognition elements are membranes, enzymes, antibodies, or microorganisms. These recognition elements are usually connected to physiochemical transducers, which in turn function due to electrochemical, optical, thermal, or piezoelectric property to convert the chemical or biological signal into measurable analytical signal. Research on chemical and biological sensors is one of the most dynamic and active research areas during the past three decades. This interest and activity is driven by the continuous need of sensors for physiological environments for human health and drug delivery. Among the different categories of sensors, electrochemical sensors are of special interest due to their suitability for miniaturization. Further, electrochemical sensors are divided in into conductometric, amperometric, and potentiometric sensors. Potentiometric sensors are suitable for physiological environments as they do not need energy source and do not rely on a passing current. Potentiometric sensors rely on the measurements of the induced electromotive voltage due to charge accumulation.

In this chapter, we will present potentiometric sensors based on ZnO nanostructures. Mainly ZnO nanowires (NWs) and nanotubes (NTs) grown on thin metallic surface or on the tip of sub-micrometer glass pipette were used as potentiometric sensors. These sensing active elements are based on ZnO nanostructures grown by the low-temperature chemical growth. The application for both extra- as well as intracellular measurements using these sensing elements to detect chemical ions and biological analytes in both intra- as well as extracellular environments will be demonstrated. One of the unique advantages of these sensors is their ability to detect relatively small concentrations in small volumes.

4.2 Growth and Fabrication of ZnO-Based Electrodes

As mentioned above, the sensing electrodes that will be presented here are based on ZnO NWs grown on either sub-micrometer glass pipettes (borosilicate glass capillaries) or thin metallic layer deposited on glass slide. For the borosilicate glass capillaries, the process starts by covering part of the capillary, including the tip by a metallic layer to be used as a contact to transfer electrical signals to the outside world. The capillary is first loaded into an evaporation system and thin-layer chromium followed by a thin layer of silver (Ag) with thicknesses of around 10 and 100 nm, respectively, is deposited on the tip. To such capillaries were prepared. One is used to process a reference electrode. For the reference electrode Ag/AgCl layer is usually used. To prepare this sub-micrometer Ag/AgCl reference electrode, one capillary with the thin silver layer is used. An electrochemical process was used to prepare Ag/AgCl. This was achieved as follows: The tip end of the silver-coated capillary was dipped into 0.2M HCl and then the Ag film was electrolyzed to form AgCl by applying 1.0 V for 1 min. By electrolyzing, a 3 cm-long layer of Ag/AgCl was deposited on the tip of the glass capillary. This layer was covered by an insulating material leaving only 3 mm at the tip uncovered. This was used a reference electrode. The other end of this reference electrode was glued to a copper wire by using high-purity silver paste and the copper wire was alter connected to the measurement setup. The working electrode was prepared from the other capillary with the Ag layer. This working electrode was prepared by growing hexagonal ZnO NWs on a small part of the tip (few millimeters). We used the low-temperature (<100°C) growth method developed in [9, 10]. In brief, in this method, first the tip of the silver-coated capillary is dipped into a seed layer solution. ZnO nanorods were grown in 150 mL of aqueous solution of 0.025 M zinc nitrate ($Zn(NO_3)_2$) and 0.025M hexamethylenetetramine (HMT, $C_6H_{12}N_4$) in a conventional flask. The reaction temperature was kept at 95°C and kept for a few hours, typically around 2–4 h. Then the working electrode with grown hexagonal ZnO NWs was removed from the growth solution and was rinsed carefully with deionized water to remove any residuals. The tip part that was covered by grown ZnO NWs was varied from 10 μm to 3 mm. The longer ones, i.e., 3 mm grown ZnO NWs tip working electrodes, were used to verify intracellular measurements (see below). Figure 4.1 shows scanning electron microscope (SEM) images of the grown ZnO NWs at different magnifications. As can be seen, the grown ZnO NWs are of hexagonal shape and cover the tip surface area densely. The working electrode with a grown ZnO NWs configuration as the example shown in Fig. 4.1 constitutes a highly sensitive probe. For the sensors developed on glass slides, the growth was performed in the same way as for the capillaries but using glass slides covered with metal layer with variations of the nutrients solution as desired [11]. The same preparation of the Ag/AgCl and the ZnO working electrode can also be applied on thin silver wires as demonstrated in [12, 13]. In the above-mentioned process, usually ZnO NWs are grown; nevertheless, other ZnO nanostructures can also be obtained and used efficiently for sensing, e.g., nanotubes, nano-flakes, and nano-walls [11, 14, 15].

Figure 4.1 Scanning electron microscope image of ZnO nanowires grown on the tip of glass capillary at different magnifications. This was used as working electrode after proper functionalization of the tip according the specific measurements.

4.3 Extra- and Intracellular Sensing of Chemical Ions

The grown working electrodes were used to measure the ionic concentration of most of the important ions that regulate important functions inside the human body. In the following text, we show some of these sensing experiments.

4.3.1 Intracellular Ca^{2+} Ion Detection

Calcium is essential in living cells. It plays an important role in regulating enzyme activity, in neural activity, in muscle contraction, cell death, etc. [16]. Hence, the determination of calcium ions (Ca^{2+}) in single cells has been of growing interest during the past few years. Different techniques relying on fluorescence spectroscopy have been used [17]. Such techniques need a large setup containing digital florescent imaging instruments. The use of the working electrodes shown in Fig. 4.1 for the detection of calcium ions or any other metallic ions can be achieved by using a much simpler experimental setup. Indeed, using ZnO NW-based functionalized working electrode together with an Ag/AgCl electrode processed from Ag coated sub-micrometer glass capillary intracellular Ca^{2+} detection has been demonstrated [16]. For these experiments borosilicate glass capillaries (sterile Femtotip II) with a tip inner diameter of about 0.5 μm, out diameter of 0.7 μm and a length of about 50 mm were used. Then ZnO NWs were grown on the tip. For these experiments two different configurations of the tip length that is covered by ZnO NWs were prepared. In one working electrode, only a very small part of the tip was covered by ZnO NWs, while in the other working electrode, a much longer tip length was used to grow ZnO NWs. The two lengths covered with ZnO NWs were 10 μm and 3 mm, respectively. These were used to verify the measurements results (see below). To access the intracellular Ca^{2+} concentration two different kinds of cells were used [16]. These were fat cells and frog cells. To functionalize the working electrodes for selectively detect only the calcium ions, 120 mg of powdered polyvinylchloride (PVC) was dissolved in 5 mL of dibutyl phthalate (DBP) and 10 mg of Ca^{2+} specific ionophore (DB18OC6). The working electrode tip was then dipped into this solution and left to dry. This process was repeated two times to insure full coverage of the selective solution over the ZnO NWs. Finally, for conditioning the working electrodes, they were dipped into 10 mM $CaCl_2$ solution [12]. Then the Ag/AgCl working electrode processed on the silver-coated glass capillary was calibrated versus a commercial Ag/AgCl.

The intracellular measurements were performed on adipocytes (fact cells) and on oocytes (frog cells). The cell preparation is described in [16]. To perform the measurements the functionalized working electrode and the Ag/AgCl reference electrodes were mounted on a micromanipulator and gently inserted into the cell under measurement. By inserting both electrodes into the cell and connecting the other two ends to a sensitive voltmeter, the electrochemical cell voltage of the activity inside the cell can be measured. The actual electrochemical potential configuration of the present system is as follows:

Ag | ZnO | buffer | | Cl^- | AgCl | Ag

Before performing the intracellular measurements, the response of the working electrode versus an Ag/AgCl was performed to obtain a calibration curve to be used to access the concentration from intracellular measurements. For this calibration, a $CaCl_2$ buffer solutions of known calcium concentrations were used. The calibration curved was then later used to access the intracellular calcium ion concentration. To validate the intracellular measurements, two different working electrodes were used. These were an electrode with a very short part of the tip covered with functionalized ZnO NWs and the other was the one with long part (3 mm) covered with ZnO functionalized NWs.

The schematic diagram shown in Fig. 4.2 shows the validity experiment setup. Using the configuration shown in Figs. 4.2a,b, intracellular measurements were performed while changing the calcium ion concentration in the buffer solution. The calcium ion concentration varied in the buffer solution outside the cell from 100 nM to 10 mM. The variation of the calcium concentration in the buffer solution was only observed for the case of the configuration shown in Fig. 4.2a. This implies that the potential measured while the whole active electrode is inserted inside the cell reflects in reality the activity inside the cell. The result of the variation of the calcium ion concentration in the buffer solution is shown in Fig. 4.3. The inset shown in this figure is the calibration curve of the working electrode used for the measurement of the intracellular calcium ion concentration. The intracellular calcium ion concentration obtained from the adipocytes cell was measured to be 123 ± 23 nM. This value is very close to the value of the intracellular concentration determined after loading human adipocytes with the Ca^{2+}-binding fluorophore fura-2 [18]. From the second experiment of the frog oocytes cells the calcium ion concentration was determined to be 250 ± 50 nM, also very close to the value reported earlier using other technique [19]. These results indicate that the technique of using ZnO nanowires as ion sensitive electrode is a promising simple and robust technique. It worth to mention that one of the most important advantages of this technique is that it can measure low concentrations in small volume (cells). This advantage will be vital for the analysis of samples for cases when the available volume to be detected is relatively small.

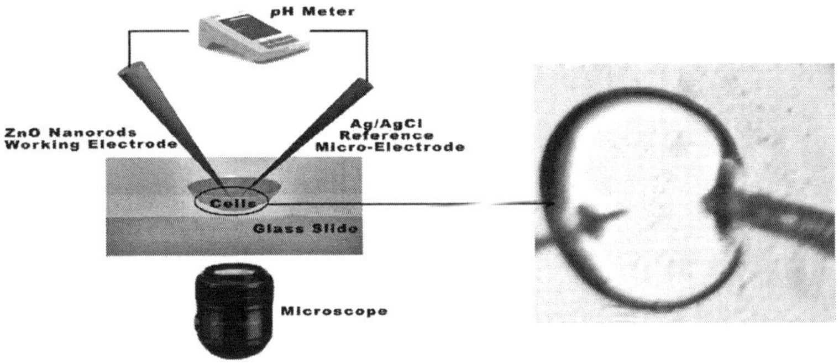

Figure 4.2 Typical experimental setup used for the intracellular Ca^{2+} concentrations measurements (left), and a digital micrograph taken while performing the intracellular measurements (right).

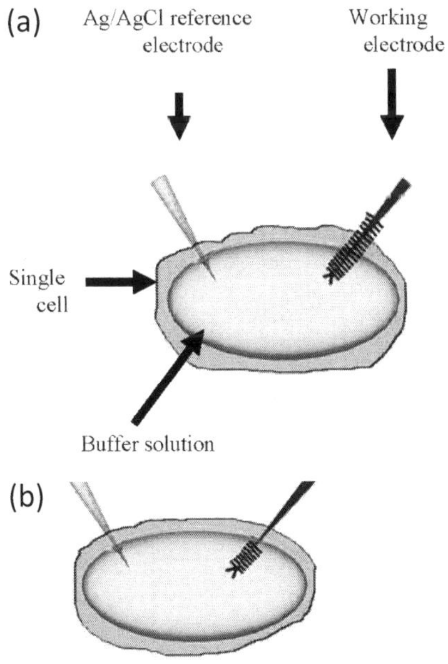

Figure 4.3 Measurement validity experimental diagram. In (a) the working electrode sensing active area is partially inserted inside the cell, while in (b) the whole active area is inserted inside the cell [16].

This selective ion detection method using ZnO NWs as the active sensing elements was also demonstrated on thin metallic wires and it showed good performance in a wide dynamic range with high sensitivity [12]. A 900 nm-long ZnO NWs of diameter 100 nm were grown on a 0.25 mm-thick Ag wire functionalized for calcium ions as described before and applied to small extracellular buffer solutions of known concentrations. The results of sensing in the buffer solutions gave results consistent with the known concentration of the calcium ion concentration of the buffer solution. The potentiometric response was found to be linear in a large logarithmic concentration range (1 μM to 0.1 M) using a standard Ag/AgCl reference electrode [12]. In addition, high sensitivity of 26.55 mV/decade was obtained. The Ag wire sensor also showed good stability [12]. It is also important to mention that these working electrodes were tested for sensing in different pH values. The results indicate that the performance is not affected at normal pH values, i.e., close to neutral pH of 7. Only at extreme pH value the performance is affected. Since all biochemical process in physiological media are at or close to neural pH, the presented approach is of potential for sensing in real physiological media.

4.3.2 Intracellular Na^{1+} Ion Detection

Another important metallic ion is Na^{1+}, which is related to many important intracellular processes and functions [20]. Sodium ions are involved in excitability of nerve and muscle cells and in water and salt homeostasis in biological systems [20]. More important, sodium ions are important for early stages of apoptotic processes, i.e., programmed cell death. Hence, it is of great interest to detect intracellular sodium ions concentration.

To functionalize the ZnO NWs working electrode, we immobilized sodium ion ionophore membrane. The best for sodium ion selectivity is the polyvinyl chloride (PVC). The composition of this PVC membrane is 1% Na^{+} ionophore ETH 227, 0.5% sodium tetraphenylborate, 75% 0-notrophenyl octyl ether, and 23% PVC. The immobilization plastic solution was prepared by dissolving in 5 mL tetrahydrofuran (THF). Then the working electrode was manually dipped into the functionalization solution and then left to dry in air. Finally, to condition the electrode for sodium ion measurements the functionalized electrode was dipped into a

10 mM NaCl solution [20]. Then buffer solutions with different sodium ions concentration range between 0.5 mM and up to 100 mM were used to obtain a calibration curve. Figure 4.4 shows the results of calibration. As can be seen, the performance was linear over the tested concentration range. These measurements showed also an excellent sensitivity of 72 mV/decade at room temperature. Again no change in the performance was observed when varying the pH around the neutral value as observed for the case of calcium ions above. Then the working electrode and the reference Ag/AgCl were used to perform intracellular measurements. We have used the same type of cells used for the calcium ion measurements, i.e., oocytes frog cells and rat adipocytes cells. The Na^{1+} concentration in adipocytes cell was found to be 11.5 mM. This value is very consistent with the value reported earlier using other technique [21]. While for the oocytes frog cells the sodium ion concentration was measured to be 8.0 mM. Again this value corresponds very closely to the value reported elsewhere [22].

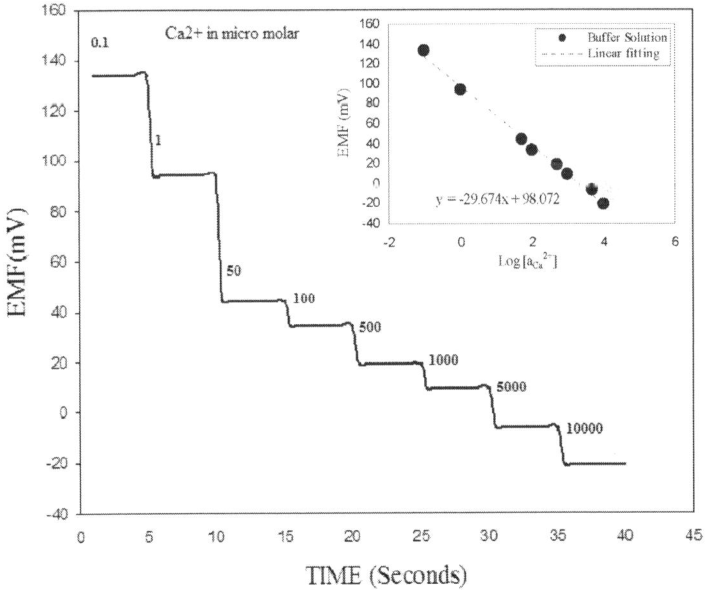

Figure 4.4 The potentiometric response versus time as the concentrations of Ca^{2+} concentration is changed in the buffer surrounding the cell for the case shown in Fig. 4.2a. The inset shows a typical the calibration curve of the present working electrode (100 nM to 10 mM) [16].

Other important metallic ions like potassium and magnesium ions were all measured in both intracellular as well as extracellular environments using the sensors configuration that relies on ZnO NWs grown on submicrometer glass capillary [23]. The results were very promising and it indicated the potential of this technique especially it suitability for sensing in small volumes, low concentration combined with the simplicity and robustness.

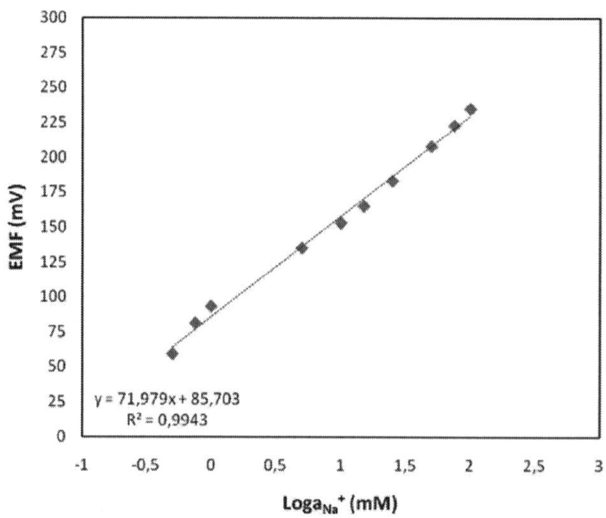

Figure 4.5 Calibration curve of using the developed ZnO NWs grown on the tip of glass capillary versus Ag/AgCl reference electrode.

4.4 Extra- and Intracellular Sensing of Glucose

Glucose, which is also known as grape sugar or corn sugar, is a fundamental carbohydrate in biology [24]. It is one of the main products of photosynthesis and serves as the human body's primary source of energy. Living cell uses it both as a source of energy and as a metabolic intermediate in the synthesis of more complex molecules such as fats. When glucose levels in the bloodstream are not properly regulated, diseases such as diabetes can develop. Because of the high demand for blood glucose monitoring, significant research has been devoted to producing reliable methods for in vitro or in vivo glucose measurement [24]. Many techniques have been developed to accurately monitor the glucose due to its important role in the human body and its effect on health. Hence, significant research has been devoted to develop reliable methods for in vitro or in vivo glucose measurement [24]. To mention few, fluorescence spectroscopy, diffraction spectroscopy, surface-enhanced Raman scattering, a wireless magneto elastic sensor, and an electrochemical transistor sensor, an enzyme-based amperometric sensor, are some of the developed methods for glucose measurements [25–29]. Nevertheless, all these techniques rely on complex equipment. A simple robust accurate technique is hence a necessity. The use of glass sub-micrometer capillary with ZnO NWs can be an excellent candidate to fulfill this necessity. Indeed, successful intracellular glucose measurements were performed using the technique described above. Again the same type of cells, i.e., adipocytes and oocytes, were used to access the intracellular glucose concentration in these two different cells.

The sensing mechanism of the electrochemical glucose sensors is based on an enzymatic reaction catalyzed by glucose oxidase (GOD) with β D-glucose, according to the following reaction:

$$H_2O + O_2 + \text{D-Glucose} \xrightarrow{\text{GOD}} \text{D-gluconolactone} + H_2O_2 \quad (4.1)$$

As a result of this reaction, D-gluconolactone and hydrogen peroxide are produced [24]. These two products and the oxygen consumption can be used for glucose determination. With H_2O availability in the reaction, gluconolactone is spontaneously converted to gluconic acid, which at neutral pH forms the charged products of gluconate$^-$ and a proton (H^+) according to the following equation:

$$\text{D-Gluconolactone} \xrightarrow{\text{Spontaneous}} \text{Gluconate}^- + H^+ \quad (4.2)$$

This proteolytic reaction of D-gluconolactone to gluconic acid shown in Eq. (4.2) results in a decrease of the medium pH that can be used for determination of the glucose concentration [30]. In our case, it is the resulting change in ionic distribution around the ZnO nanorods that causes a change of the overall potential of the ZnO-nanorod electrode [24].

To functionalize the working electrodes, glucose oxidase is hence used. Figure 4.6 shows a micrograph of the two cells under glucose measurement. Before the intracellular measurements, a calibration curve was obtained. Figure 4.7 shows the calibration curve obtained.

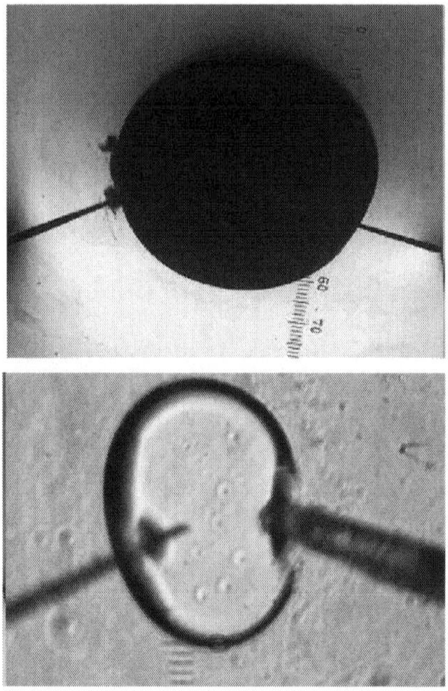

Figure 4.6 Micrograph images of the two cells adipocytes and oocytes under intracellular glucose measurements. (up) The oocytes cell, and (down) adipocytes cell.

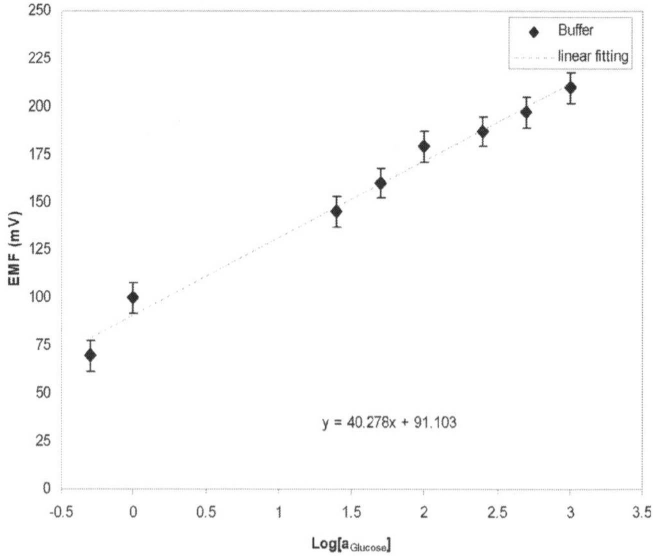

Figure 4.7 Typical calibration curve obtained for glucose assessment.

The values of the intracellular glucose concentration were again consistent with the values obtained using other techniques; for more details, see Ref. [24]. Moreover, the insulin-induced activation of glucose uptake was also investigated. Figure 4.8 shows the diagram of the intracellular mechanism for insulin-induced activation glucose uptake. Figure 4.8b shows the output response with respect to the time of the working electrode when insulin is applied to the external environment around the cell [24].

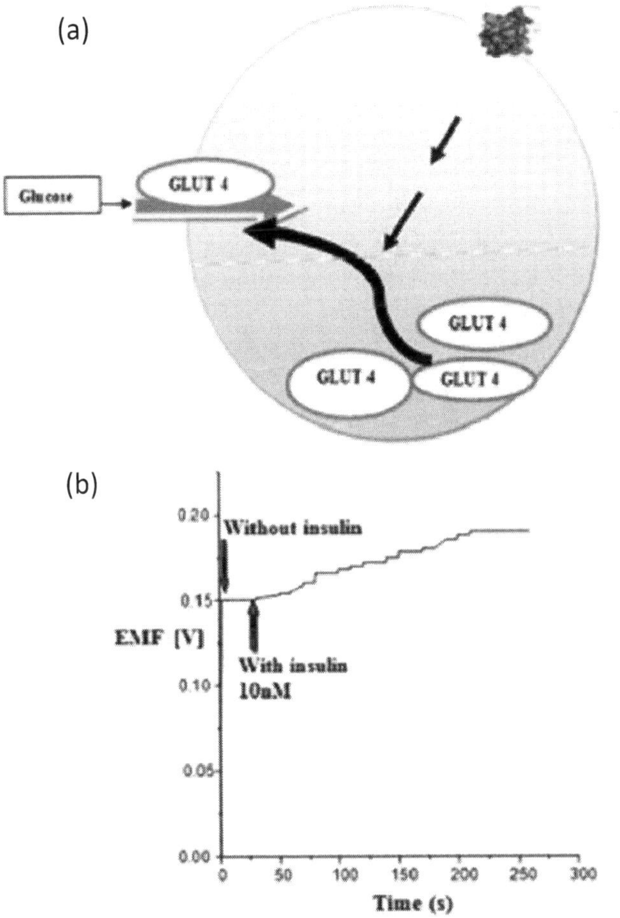

Figure 4.8 (a) Intracellular mechanisms for insulin-induced activation of glucose uptake, and in (b) the working electrode response with and without insulin around the external environment around the cell [24].

4.5 Extra- and Intracellular Sensing of Biological Analytes and Other Chemical Substances

The technique of using ZnO NWs as sensitive selective electrodes grown on the tip of a sub-micrometer glass pipette has also been used to estimate the intracellular as well as the extracellular amounts of different biological analytes and other chemical substances. Below we give two examples for the detection of cholesterol and uric acid as the concentration of both in the human body is important.

4.5.1 Cholesterol Sensing Using ZnO Nanostructures

Cholesterol is a building block for the human body tissue and is an essential element in the cell membrane [31]. Metabolisms of cholesterol have many important biological roles such as in the production of bile acid, in sex hormones, in signal transduction, and in solubilizing body lipids [32, 33]. Nevertheless, excess of cholesterol in the human body is one of the main causes in the pathogenesis of cardiac and brain vascular diseases and is also implicated in dementias, diabetes, and cancer, as well as in several rare monogenic diseases [34–36]. Hence, balanced levels of cholesterol in human body are essential for healthy life. This urges the need for a simple, sensitive, selective, robust method of measuring the cholesterol in human body to monitor its level.

Low-temperature chemical growth was used to grow ZnO NWs on the tip of a thin (80.25 mm) silver wire. To tune the selectivity of the ZnO NWs and functionalize it for the monitoring of cholesterol oxidase was used [31]. For biosensors, the use of enzymes for selectivity is very common since they can catalyze chemical reaction as required. Cholesterol oxidase (ChOx), which is a key enzyme in cholesterol metabolism, is the best enzyme to choose for tuning the selectivity for cholesterol detection. Combing ZnO NWs and ChOx is also efficient since the isoelectric points for the two is 9.5 and 4.5, respectively [31]. The large difference in the isoelectric points between ZnO and ChOx makes the immobilization of ChOx on the surface of ZnO NWs easily achieved by simple electrostatic interaction [31]. Figure 4.9 shows different SEM images of the ZnO NWs at different magnifications with or without cholesterol on their surface.

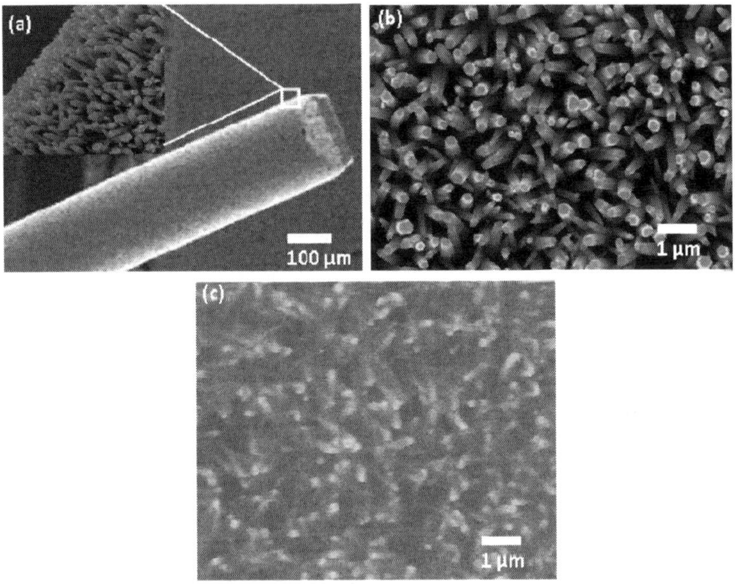

Figure 4.9 (a) Low magnified SEM image of Ag wire containing ZnO nanorods. Inset shows vertically grown ZnO nanorods on the small portion of Ag wire. High-magnification SEM image of (b) as-grown ZnO nanorods before exposure to cholesterol test solution. (c) ZnO nanorods having ChOx film after the exposure to the cholesterol test solution.

The setup of the cholesterol measurements using ZnO NWs grown on Ag thin wire is shown in Fig. 4.10.

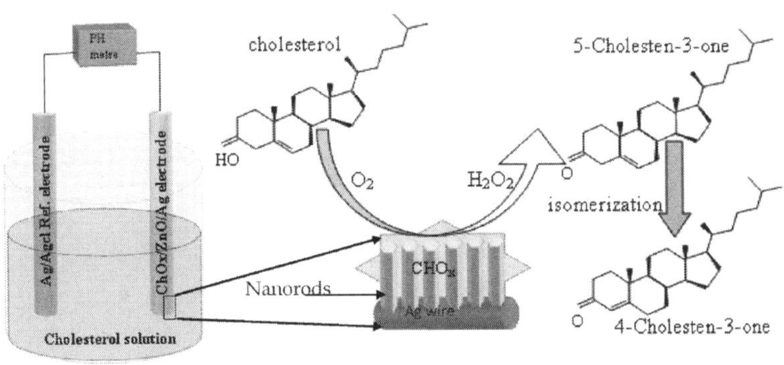

Figure 4.10 Schematic illustration of the cholesterol sensing setup using working bio-electrode comprising ZnO nanorods, cholesterol oxidase (ChOx), and Ag wire, along with possible electrochemical reaction near the working electrode [31].

After electrostatic immobilization of ChOx on the surface of the ZnO NWs, cholesterol electrolyte solutions with different concentrations were prepared. The concentration range investigated was between 0.001 and 10 mM. The electromotive-induced potential was measured with respect to commercial Ag/AgCl reference electrode and variation in this induced electromotive potential was observed upon changing the electrolyte solution cholesterol concentration. The measured value corresponds very closely with the known control values of the prepared electrolyte cholesterol solution. This ZnO NW–based cholesterol biosensor shows a relatively high sensitivity of about 35 mV/decade [31]. Reproducibility experiments were also conducted for the present cholesterol biosensor and the results indicated that this biosensor possesses acceptable reproducibility [31]. It is also important that using other ZnO nanostructures, it was possible to construct more sensitive cholesterol biosensors. Sensitivity with a value of 53 mV/decade was demonstrated for cholesterol sensing using ZnO Nano flakes (nanowalls) of with thickness of around 80 nm [37]. One reason for the improved sensitivity when comparing nano flakes to nanowires is the increased surface-area-to-volume ratio.

4.5.2 Uric Acid Detection Using ZnO Nanowires

On the other hand, uric acid is a major nitrogenous compound in the urine, is a product of purine metabolism in the human body, and is related to much clinical disorder [38]. Hence, it is important to determine the concentration of uric acid in the urine and in blood samples using a fast, robust, and sensitive simple technique. Zinc oxide NWs grown flexible plastic substrate were functionalized with uricase, which gives excellent selectivity for uric acid [39]. To prepare the selective electrode, a 10 mM phosphorous buffered solution (PBS) with a pH 7.4 solution was used by dipping the ZnO NWs electrode. After dipping the electrode in the enzyme solution it was left to dry in air. Finally, to reduce any possible leakage from the enzyme-coated electrode and avoid interference from other species, a 5 μL Nafion® solution (1% in methanol) was applied to the electrode [39]. The functionalized ZnO NWs electrode was then used versus a commercial Ag/AgCl electrode. The uric acid biosensor was then tested in standard known solutions of uric acid. The response of this biosensor is shown in Fig. 4.11. Then the uric acid biosensor was checked for reproducibility and stability. Figure 4.5 shows the response of the same electrode from three different experiments. After each experiment the electrode was soaked carefully in PBS, and then it was reused for sensing of uric acid. As shown in Fig. 4.12 the electrode shows excellent reproducibility.

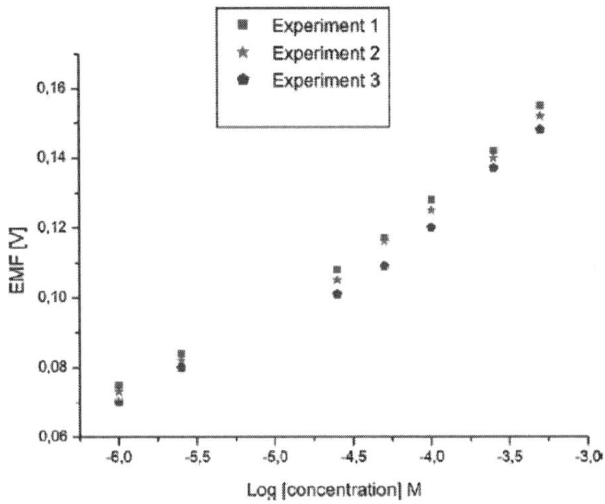

Figure 4.11 The potentiometric response of the ZnO NW–based biosensor from three different experiments indicating excellent reproducibility.

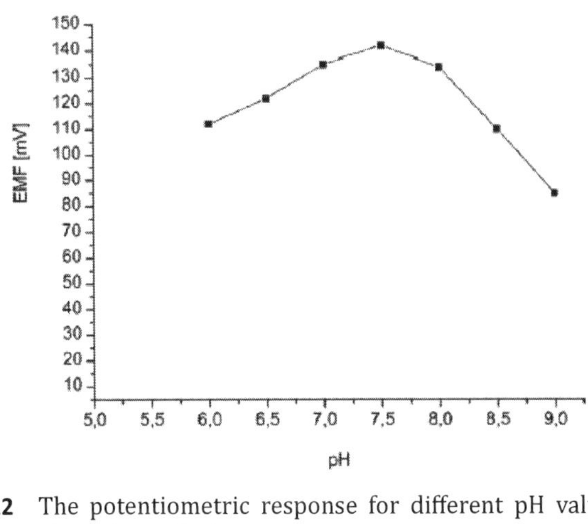

Figure 4.12 The potentiometric response for different pH values of the electrolyte solution.

Further, the effect of measurements temperature and solution pH on the uric acid biosensor was both investigated. The performance was investigated over a pH range between 6 and 9. It was found that the sensors performance this uric acid biosensor was affect by pH. Figure 4.12 shows the performance of the biosensor in a 250 μM uric acid solution; the figure indicates that the optimum pH operating condition of this biosensor is at a pH range between 6.5 and 7.5. Since the pH of most of the physiological environments is within this range, the result shown in Fig. 4.12 indicates that uric acid biosensor based on ZnO NWs is suitable for human uric acid determination [39].

Further, the effect of temperature on the performance of the uric acid biosensor was also investigated. The temperature of the uric acid solution was varied between 20 and 70°C [39]. As seen in Fig. 4.13, as the temperature is raised from 20°C, the electromotive potential starts to increase. This increase reaches a maximum at around 40°C, then by further increasing the temperature, the electromotive potential starts to decrease. Although room temperature is not the optimum temperature for operating this biosensor, it is advised to operate it at room temperature to reduce any possible evaporation effects [39]. Finally, interfering elements effects on the performance of this uric acid biosensor were investigated. Due to the specificity of the uricase enzyme and its high selectivity to uric acid, negligible interference effects were observed [39].

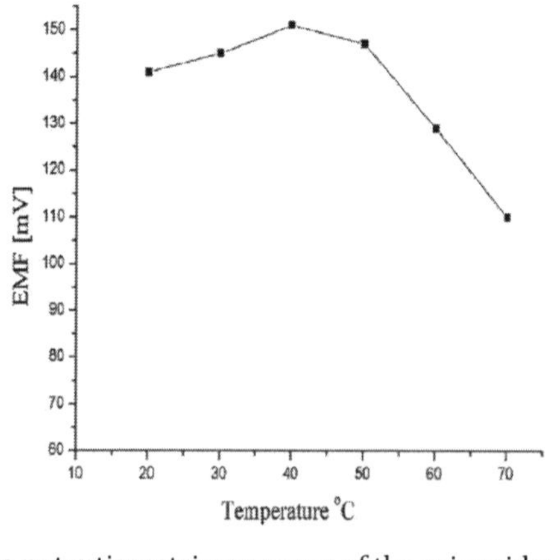

Figure 4.13 The potentiometric response of the uric acid when varying the temperature.

4.6 Zinc Oxide Nanowires as Research Tool for Cancer Cell Treatment

Photodynamic therapy (PDT) has become a popular method for treatment of cancer cells [40, 41].

Photodynamic therapy in cancer cells involves the uptake of a photosensitizer by a cancer cell followed by exposure to white light [42]. The choice of a suitable irradiation wavelength is one of the basic parameters affecting the efficiency of the PDT. Irradiation with long wavelength light to protoporphyrin IX dimethyl ester (PPDME), which has absorption at a wavelength of 630 nm of light, usually provides low quantum yields for singlet oxygen [43]. Usually the photosensitizer is applied on the tumor for a typical duration of 3–6 h before light exposure so that an adequate amount of it can penetrate into the tumor [42]. The molecular oxygen in the cell is essential for the PDT, as the phototoxic reaction involves the formation of singlet oxygen in the PDT [44]. The singlet oxygen is produced by the energy transfer between the molecular oxygen and the triplet state of excited photosensitizer molecules in the cell [45]. This singlet oxygen will cause initial damage to the mitochondria leading to cell death [22]. Figure 4.14 shows a schematic diagram illustrating the photodynamic therapy process inside the cell. The cell death (necrosis) due to the damage of the mitochondria is still under debate. However, the penetration of drug (aminolevulinic acid (ALA)) is the major limiting factor in PDT [46]. For this reason, there is demand for new and improved photosensitizer delivery systems for effective treatment modalities. The ability of ZnO to emit white light [2] and its success in when grown on sub-micrometer as safe intracellular invasive tools has encouraged the use of ZnO NWs as intracellular photosensitizer carriers.

Zinc oxide NWs grown as described above on the tip of a submicrometer glass capillary were sued for PDT experiments [42]. These ZnO NWs were then conjugated with PPDME and inserted inside cancerous cell. Figure 4.15 shows digital micrographs obtained during PDT experiments. In Fig. 4.15a, the conjugated tip is shown before entering the cell, while Fig. 4.15b shows the tip inserted inside the cancer cell.

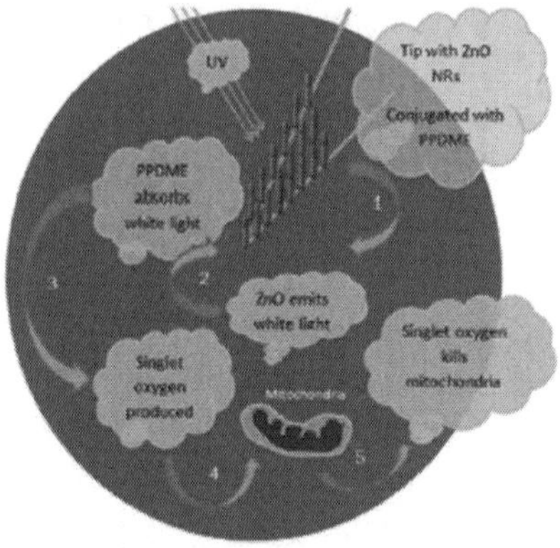

Figure 4.14 Schematic diagram illustrating the photodynamic therapy process inside the cell. Digital images taken during the performance of the intracellular measurements, (a) image of the PPDME conjugated ZnO NWs tip before entering the cell, (b) image of the PPDME conjugated ZnO NWs tip inside the cancer cell, and in (c) conjugated ZnO NRs tip after cancer cell necrosis

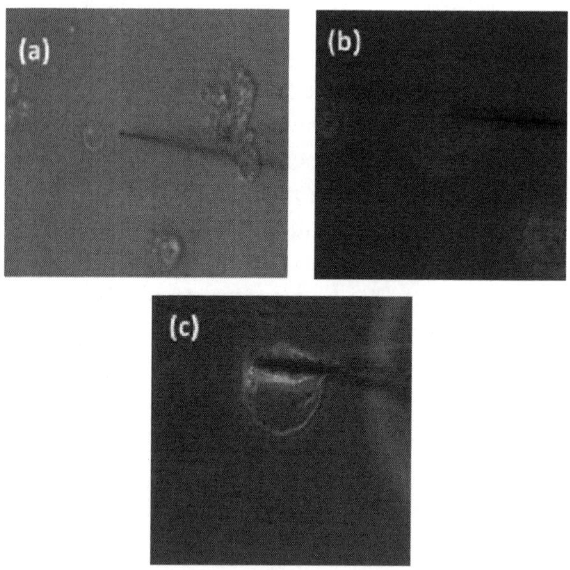

Figure 4.15 Digital images taken during the performance of the intracellular measurements, (a) image of the PPDME conjugated ZnO NWs tip before entering the cell, (b) image of the PPDME conjugated ZnO NWs tip inside the cancer cell, and in (c) conjugated ZnO NRs tip after cancer cell necrosis.

After the insertion of the conjugated ZnO NWs tip, UV was applied for 10 min. As observed, tumor cells are rapidly dividing cells with a destroyed outer barrier (photograph not shown). After the UV exposure the cell morphology as shown in Fig. 4.15c indicates cell necrosis death. These experiments were also performed with bare ZnO NWs, and upon exposure to UV for 10 min, no cell necrosis was observed [42].

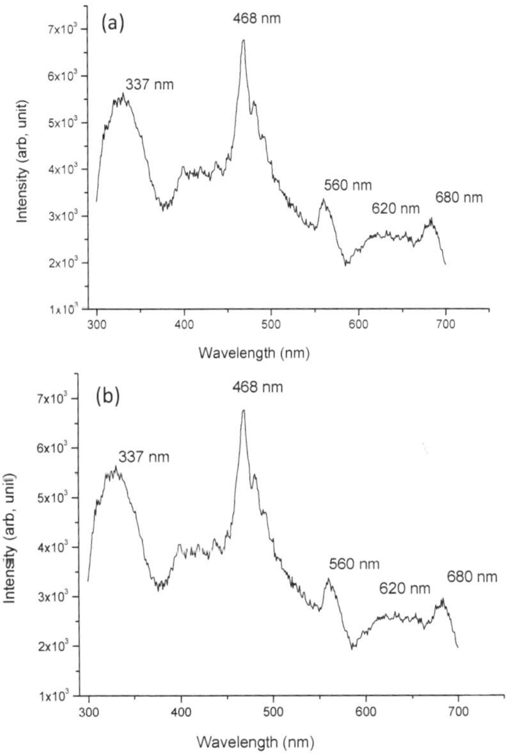

Figure 4.16 Fluorescence spectra of ZnO NWs (a) without and (b) with, protoporphyrin, respectively. These spectra were recorded using a laser of wavelength 240 nm as the excitation source using a PTI Fluorescence system.

To identify the emission of PPDME conjugated ZnO NWs, extra experiments can be carried out [42]. Exciting different ZnO NWs, with and without conjugation with PPDME can reveal useful information. Such experiment has been performed and yielded the spectra shown in Fig. 4.16. In Fig. 4.16a, the fluorescence (Fl) spectrum shows five different peaks. The origin of these peaks is discussed in details in [42]. Figure 4.16b shows the fluorescence emission of the PDT conjugated ZnO NRs device. Basically, it is consisting of the same peaks, except a small chănge of the nature due probably the absorption by the PPDME [42]. Moreover, the intensity of the Fl in Fig. 4.16b is five times less than that of the Fl spectrum in Fig. 4.16a indicating absorption of some of the emitted white light by the PPDME as expected. From the combined results shown in Figs. 1.6.3 and 4.15, it is concluded that white light emission from ZnO NWs was transferred by absorption to the PPDME, which in turn caused excitation and production of singlet oxygen, which consequently caused the aimed cell necrosis.

4.7 Interfacing ZnO Nanowire–Based Sensors with Conventional Electronic Components

Usually all sensors are connected to conventional electronic components, e.g., transistors in order to be to amplify the signal and hence obtain a readable information. The above described ZnO NWs sensors would be useless if the signal obtained from them is too weak to be interfaced to conventional electronic circuits. Nevertheless, results of coupling ZnO NW–based sensors indicate that the signal is strong enough to be detected by conventional electronic components for amplification, i.e., transistors [47, 48].

Using ZnO NWs grown on Ag thin wire, the coupling to conventional electronics was demonstrated for the sensing of calcium ions and glucose [47, 48]. A schematic diagram showing the configuration using when coupling ZnO NWs on Ag wire to a commercial transistor is shown in Fig. 4.17. For the calcium ions, by coupling the Ag wire with functionalized ZnO NWs directly to the gate of a commercial metal oxide field effect transistor (MOSFET), the electrochemical effect of varying the ions concentration in an electrolyte control solution was inducing a clearly measurable effect on the gate of the transistor [47]. The same effect was also observed for the case of glucose detection using extended gate commercial MOSFET [48]. Finally, we demonstrate the interfacing of the sensor to mobile telephone system for signal transmission [49].

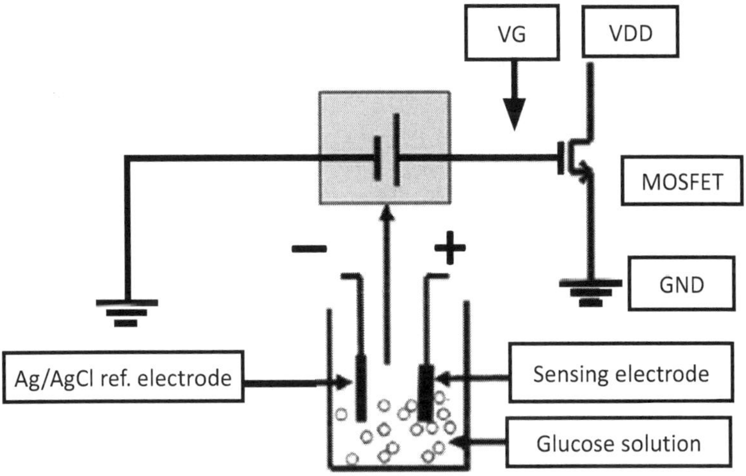

Figure 4.17 Schematic diagram illustrating the configuration used for glucose detection with MOSFET using extended gate functionalized ZnO nanowires as working electrode and Ag/AgCl as a reference electrode.

4.8 Summary

In summary, ZnO is biosafe and biocompatible and has excellent electron transfer properties, which implies a potential for the development of sensing devices. The fact that it also possesses self-organized growth property with a possibility of wet chemical growth at relatively low temperature enables the growth on soft glass capillaries. This implies that ZnO nanostructures, and especially nanowires, with relatively large surface-area-to-volume ratio can be utilized for sensing devices capable of detecting low concentrations in small volumes. The ability to sense in small volumes is indeed an important advantage, as sometimes the available volume for sensing is already relatively small, e.g., in crime scenes. The use of ZnO NWs grown on the tip of a sub-micrometer glass capillary has proven to be a very efficient approach to develop highly sensitive sensors for detection of chemical ions and other biological analytes. Moreover, ZnO NWs with their ability to emit white light have proved to also be efficient for local photodynamic therapy as a research tool, i.e., as photosensitizer carrier system. Finally, we were able to interface the sensor to the mobile telephone system.

References

1. Klingshirn, C. (2007). ZnO, Material, physics and applications, *Chem. Phys. Chem.*, **8**(6), 782–803.
2. Willander, M., Nur, O., Zhao, Q.X., Yang, L.L., Lorenz, M., Cao, B.Q., Zúñiga Pérez, J., Czekalla, C., Zimmermann, G., Grundmann, M., Bakin, A., Behrends, A., Al-Suleiman, M., El-Shaer, A., Che Mofor, A., Postels, B., Waag, A., Boukos, N., Travlos, A., Kwack, H.S., Guinard, J., Le Si Dang, D. (2009). Zinc oxide nanorods based photonic devices: recent progress in growth, light emitting diodes and lasers, *Nanotechnology*, **20**, 33201.
3. Sheini, J.F., Joang, D.S., More, M.A., Singh, J., Srivasatva, O.N. (2010). Low temperature growth of ZnO nanowires and their application as field emission cathodes, *Mater. Phys. Chem.*, **120**, 691–695.
4. Zang, Z.L., Song, J. (2006). Piezoelectric nanogenerators based on zinc oxide nanowire arrays, *Science*, **321**, 242–246.
5. Li, Z., Yang, R., Yu, M., Bai, F., Li, C., Wang, Z.L. (2008). Cellular level biocompatibility and biosafety of ZnO nanowires, *J. Phys. Chem. C*, **112**(51), 20114–20117.
6. Brochen, S., Granier, C., Feuillet, G., Pernot, J. (2012). Role of deep and shallow donor levels on n-type conductivity of hydrothermal ZnO, *Appl. Phys. Lett.*, **100**(5), 052115(3p).
7. Finch, G. I., Wilman, H. (1934). The lattice dimensions of zinc oxide, *J. Chem. Soc.*, 751–754.
8. Willander, M., Risveden, K., Danielsson, B., Nur, O. (2009). Trapping and detection of single molecules in water, in *Micro and Nano Technologies for Bio-Analysis* (Lee, J.W., Foote, R.S., ed.), Humana Press, Chapter 12, pp. 163–186.
9. Vayssieres, L., Keis, K., Lindquist, S.E., Hagfeldt, A. (2001). Purpose built anisotropic metal oxide material: 3D highly oriented microarrays of ZnO, *J. Phys. Chem. B*, **105** (17), 3350–3352.
10. Greene, L.E., Law, M., Goldberger, J., Kim, F., Johnson, J.C., Zhang, Y., Saykally, R.J., Yang, P. (2005). General route to vertical ZnO nanowire arrays using textured ZnO seeds, *Nano Lett.*, **5** (7), 1231–1236.
11. Fulati, A., Usman Ali, S.M., Muhammad, R., Amin, G., Nur, O., Willander, M. (2009). Miniaturized pH sensors based on zinc oxide nanotubes/nanorods, *Sensors*, **9**, 8911–8923.
12. Asif, M.H., Nur, O., Willander, M. Yakovleva, M., Danielsson, B. (2008). *Res. Lett. Nanotechnol.*, **2008**, 1–4p.

13. Israr, M.Q., Sadaf, J.R., Asif, M.H., Nur, O., Willander, M. (2010). Potentiometric cholesterol biosensor based on ZnO nanorods chemically grown on Ag wire, *Thin Solid Films*, **59**, 1106–1109.

14. Fulati, A., Usman Ali, S.M., Asif, M.H., ul Hasan Alvi, N., Willander, M., Brännmark, C., Strålfors, P. Börjesson, S.I., Elinder, F. (2009). An intracellular glucose biosensor based on ZnO nano-flake, *Sens. Actuators: Chem. B*, **150**, 673–680.

15. Israr, M.Q., Sadaf, J.R., Nur, O., Willander, M., Salman, S., Danielsson, B. (2011). Chemically fashioned ZnO nano-walls and their potential application for potentiometric sensing, *Appl. Phys. Lett.*, **98**, 253705.

16. Asif, M.H., Fulati, A., Nur, O., Willander, M., Brännmark, C., Strålfors, P., Börjesson S.I., Elinder, F. (2009). Functionalized zinc oxide nanorod with ionophore-membrane coating as an intracellular Ca^{2+} selective sensors, *Appl. Phys. Lett.*, **95**, 023703.

17. Nguyen, T., Rosenzweig, Z. (2002). Calcium ion fluorescence detection using liposomes containing Alexa-labeled calmodulin, *Anal. Bioanal. Chem.*, **374**, 1431–1434.

18. Draznin, B., Sussman, K.E., Echel, R.H., Kao, M., Yosti, T., Sherman, N.A. (1988). Possible role of cytosolic free calcium concentrations in mediating insulin resistance of obesity and hyperinsulinemia, *J. Clin. Invest.*, **82**, 1848–1852.

19. Takahashi, T. Neher, E., Sakmann, B. (1987). Rat brain serotonin receptors in Xenopus oocytes are coupled by intracellular calcium to endogenous channels, *PNAS*, **84**, 5063–5067.

20. Asif, M.H., Nur, O., Willander, M., Strålfors, P., Brännmark, C., Elinder, F., Englund, EU.H., Lu, J., Hultman, L. (2010). Growth and structure of ZnO nanorods on a sub-micrometre glass pipette and their application as intracellular potentiometric selective ion sensors, *Materials*, **2010**(3), 4657–4667.

21. Resh, M.D., Nemenoff, R.A., Guidotti, G. (1980). Insulin simulation of (Na^+, K^+) adenosine triphosphatase-dependant $86Rb^+$ uptake in rat adipocytes, *J. Bio. Chem.*, **255**, 10938–10945.

22. Dascal, N. (1987). The use of Xenopus oocytes for the study of ion channels, *CRC Crit. Rev. Biochem.*, **22**, 317–387.

23. Asif, M.H. (2011). Zinc oxide nanostructure based electrochemical sensors and drug delivery to intracellular environments. Linköping Studies of Science and Technology PhD thesis no. 1376.

24. Asif, M.H., Usman Ali, S.M., Nur, O., Willander, M., Brännmark, C., Strålfors, P., Englund, U.H., Elinder, F., Danielsson, B. (2010). Functionalized ZnO-

nanorods-based selective electrochemical sensor for intracellular glucose. *Biosens. Bioelectron.*, **25**, 2205–2211.

25. Ballerstadt, R., Schultz, J.S. (2000). A fluorescence affinity hollow fiber sensor based for continuous transdermal glucose monitoring, *Anal. Chem.*, **72** (17), 4185–4192.
26. Asher, S.A., Alexeev, V.L., Goponenko, A.V., Sharma, A.C., Lednev, I.K., Wilcox, C.S., Finegold, D.N. (2003). Photonic crystal carbohydrate sensor: low ionic sugar concentration sensor, *J. Am. Chem. Soc.*, **125**(11), 3322–3329.
27. Shafer-Peltier, K. E., Haynes, C. L., Glucksberg, M. R., Van Duyne, R. P., (2003). *J. Am. Chem. Soc.*, **125** (15), 588–593.
28. Cai, Q.Y., Zeng, K.F., Ruan, C.M., Desai, T.A., Grimes, C.A. (2004). A wireless, remote query glucose biosensor based on a pH-sensitive polymer, *Anal. Chem.*, **76** (14), 4038–4043.
29. Forzani, E.S., Zhang, H.Q., Nagahara, L.A., Amlani, I., Tsui, R., Tao, N.G. (2004). A conducting polymer nanojunction sensor for glucose detection, *Nano Lett.*, **4** (9), 1785–1788.
30. Shaw, G.W., Clarement D.J., Pickup, J.C., Bergveld, P. (2003). Highly sensitive glucose sensor based on work function changes measured by an E MOSFET, *Anal.*, **128**, 1062–1066.
31. Israr, M.Q., Sadaf, J.R., Asif, M.H., Nur, O., Willander, M., Danielsson, B. (2010). Potentiometric cholesterol biosensor based on ZnO nanorods chemically grown on Ag wire, *Thin Solid Films*, **519**(3), 1106–1109.
32. Issacsohn, J, in: *Heart Book*, New York, 1992, Chapter 4, 37.
33. Ikonen, E. (2008). Cellular cholesterol trafficking and compartmentalization, *Nat. Rev. Mol. Cell Biol.*, **9**, 125–138.
34. Staels, B. (2002). Cardiovascular biology: a cholesterol tether, *Nature*, **417**, 699–701.
35. Maxfield, F.R., Tabas, I. (2005). Role of cholesterol and lipid organization in disease, *Nature*, **438**, 612–621.
36. Ikonen, E. (2006). Mechanisms of cellular cholesterol transport: defects and human disease, *Physiol. Rev.*, **86**, 1237–1261.
37. Israr, M.Q., Sadaf, J.R., Nur, O., Willander, M., Salman, S., Danielsson, B. (2011). Chemically fashioned ZnO nanowalls and their potentiometric application for potentiometric cholesterol biosensor, *Appl. Phys. Lett.*, **98**, 253705–253703p.
38. Wootton, I.D.P., Freeman, H. (1982). *Microanalysis in Medical Biochemistry*, 6th ed. New York.

39. Usman Ali, S.M., Alvi, N.H., Ibupoto, Z., Nur, O., Willander, M., Danielsson, B. (2011). Selective potentiometric determination of uric acid with uricase immobilized on ZnO nanowires, *Sens. Actuators B: Chem.*, **152**, 241–247.

40. Wang, J., Yi, J. (2008). Cancer cell killing via ROS: to increase or decrease that is the question, *Can. Bio. Ther.*, **7**, 1875–1884.

41. Anand, S., Honari, G., Hasan, T., Elson, P., Maytin, E.V. (2009). Low-dose methotrexate enhances aminolevulinate-based photodynamic therapy in skin carcinoma cells in vitro and in vivo, *Clin. Can. Res.*, **15**, 3333–3343.

42. Kishwar, S., Asif, M. H., Nur, O., Willander, M., Larsson, P.-O. (2010). Intracellular ZnO nanorods conjugated with protoporphyrin for local mediated photochemistry and efficient treatment of single cancer cell, *Nanoscale Res. Lett.*, **5**, 957–960.

43. Pavel, K., Zdenek, Z., Milan, J. (1997). The effect of the irradiation wavelength on the processes synthesized Protoporphyrin IX Dimethyl Ester, *Radiat. Res.*, **148**, 382–385.

44. Bellnier, D. A., Wood, L. M., Potter, W. R., Weishaupt, K. R., Oseroff, A. R. (1999). Design and construction of a light-delivery system for photodynamic therapy, *Med. Phys.*, **26**, 1552–1558.

45. Foote, C. S. (1968). Mechanisms of photosensitized oxidation, *Science*, **162**, 963–970.

46. Morton, C. A., MacKie, R. M., Whitehurst, C., Moore, J. V., McColl, J. H. (1998). Photodynamic therapy for basal cell carcinoma: effect of tumour thickness and duration of photosensitizer application on response, *Arch Dermatol.*, **134**, 248–249.

47. Asif, M. H., Nur, O., Willander, M., Danielsson, B. (2009). Selective calcium ion detection with functionalized ZnO nanorods-extended gate MOSFET, *Biosens. Bioelectron.*, **24**, 3379–3382.

48. Usman Ali, S. M., Nur, O., Willander, M., Danielsson, B. (2009). Glucose detection with a commercial MOSFET using a ZnO nanowire extended gate, *IEEE Trans. Nano*, **8**(6), 678–683.

49. Usman Ali, S. M., Tasuif, A., Axelsson, K., Nur, O., Willander, M. (2011). Wireless remote monitoring of glucose using functionalized ZnO nanowires as based sensor, *Sensors*, **11**, 8485–8496.

Chapter 5

ZnO Nanostructures: Toxicity and Phototoxicity Characteristics in Biological Samples

Magnus Willander,[a] O. Nur,[a] M. Fakhr-e-Alam,[b] M. Atif,[c] and M. S. AlSalhi[c]

[a]*Department of Science and Technology, Campus Norrköping, Linköping University, SE-601 74 Norrköping, Sweden*
[b]*Department of Physics, Faculty of Basic and Applied Sciences, International Islamic University, Islamabad, Pakistan*
[c]*King Saud University, Riyadh, Kingdom of Saudi Arabia*

magwi@itn.liu.se

Zinc oxide nanoparticles (ZnO NPs) are emerging milestones in the ongoing research in photodynamic therapy (PDT) with tremendous multiple clinical applications, diagnostic as well as anti-tumoricidal, in many microbial non-microbial treatments and are front-runners in these applications due to their high quantum yield, size-dependent tunable emission of wavelength over a wide spectrum of light. Currently, the nano-dependent PDT technique involving ZnO NPs and zinc oxide nanorods (ZnO NRs) is simple bio-safe, biocompatible in dark, enhances endogenous fluorescence, noninvasive, and fast with their least permeability in normal cells,

Zinc Oxide Nanostructures: Advances and Applications
Edited by Magnus Willander

ISBN 978-981-4411-33-2 (Hardcover), 978-981-4411-34-9 (eBook)
www.panstanford.com

but ZnO NRs with high surface-to-volume ratio and biocompatibility can be used as an efficient photosensitizer carrier system and at the same time provide intrinsic white light needed to achieve cancer cell necrosis. In this current work, the authors demonstrate the toxicity of different ZnO nanostructures alone and complex with different photosensitizers (PS), e.g., aminolevulinic acid (ALA), Photofrin® and protoporphyrin dimethyl ester (PPDME) by using multiple malignant cell lines, e.g., hepatocellular carcinoma (HepG2), rhabdomyosarcoma (RD), cervical (HeLa), melanoma (FM55P), and foreskin fibroblast (AG1518) cells by applying multiple/various techniques. In addition, ZnO NRs are prominent semiconducting and piezoelectric materials that have multiple applications in the field of optoelectronics, biosensors, resonators, electric nanogenerators, energy scavenging, and nanolasers [1–7]. Zinc oxide (ZnO) with its semiconducting and piezoelectric properties exhibits bio-safety and biocompatibility [8]. ZnO being a wide band gap (3.37 eV) semiconductor, has large excitation binding energy (≈60 meV) at room temperature, have applications in optronics [9, 10]. Moreover, because of its electronic and optical properties ZnO nanostructures, e.g., nanorods (NRs), nanotubes (NTs), nanoparticles (NPs), and nanowires (NWs) are dominantly found in transparent electronics devices [11]. ZnO nanostructures have attracted the attention of researchers because of their potential applications in nanodevices, e.g., nanobiotechnology one-dimensional nanostructures such as nanotubes, nanowires, and nanoribbons [11–13]. In addition, the role of nanoparticles in biomedical applications cannot be ignored, e.g., targeted drug delivery, hyperthermic cancer treatment, gene therapy, ultra-sensitive bio-agent detection and magnetic resonance imaging (MRI) and overcoming of multidrug resistance [14]. In this regard, different drug delivery systems were introduced for the assessment of ZnO toxicity in dark as well as under exposure of 240 nm of UV light. Some of them are briefly discussed here:

- freestanding drug delivery
- microinjection drug delivery

Different malignant cell lines/biological samples, such as melanoma, foreskin fibroblast, breast carcinoma, hepatocellular carcinoma, and RD, were used as in vitro experimental model due to corresponding experimental equipment facility/availability. In this current conducting experiment, toxicity of different nanomaterial structures, such as NRs, nanoflakes (NFs), NPs, NPS, and NWs, were analyzed for mentioned cell lines. It was investigated that ZnO NRs have marvelous toxicity even in the absence of laser light (in dark). Various nanomaterial structures are shown in Fig. 5.1.

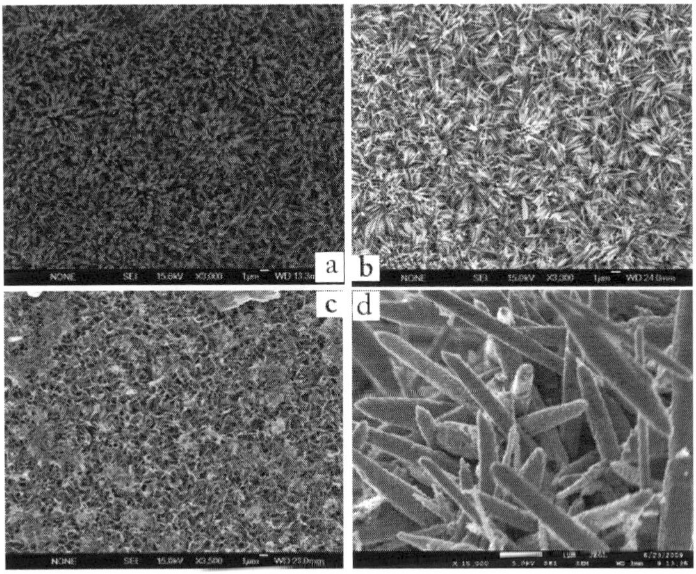

Figure 5.1 (a) Zinc oxide nanowires, (b) zinc oxide (ZnO) nano-urchin, (c) zinc oxide nanoparticles, and (d) zinc oxide nanorods.

Many research experts agreed that zinc oxide nanomaterials, in the presence of 10–20 J/cm^2 of UV light doses, showed the significant amount of reactive oxygen species (ROS) which is responsible for the cell killing effect [15]. It has been investigated that dimension and size of nanomaterials is also dependent factor for cell toxicity, sometimes significant loss in cellular viability were recorded due to shape and dimension of ZnO NPs. In multiple publicized data, it was found that the concentration of nanomaterials toward the malignant cell line has a direct relationship with the killing of the mentioned cells via cell membrane rupturing, lysosomal injury/vascular blockade [16]. In this chapter, we focus the killing of cells via photochemical reaction instead of mechanical stress/trauma, which includes in PDT drawback. The schematic diagram of photochemical reactions (Jablonski diagram) that lead to cell death/cell apoptosis or necrosis is shown in Fig. 5.2.

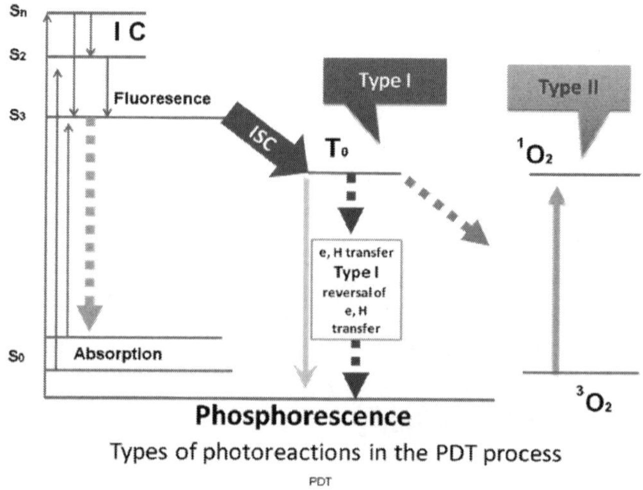

Figure 5.2 Jablonski diagram.

Different energy levels are shown by S_0, S_1, S_2, and S_3. S_0 is the lowest ground state (which is stable). S_1, S_2, and S_3 are first, second, and third excited states, respectively. When light of suitable energy (threshold dose) is to be shine to drug/chromophore if the light having property ≥ threshold dose (energy gap). The photosensitizing molecule has the probability of going to one of the excited states, e.g., S_1, S_2, and S_3. After some time, the atoms/molecules de-excite to the ground state by decaying fluorescence, which is important for the diagnostic point of view or atoms/chromophore back to triplet state (forbidden) by radiationless decay (heat). During this decay process in the second step, the energy transferred to triplet oxygen occurs in the surrounding of the targeted tissue. This may cause the production of ROS, which leads to cell death. Currently, ZnO nanoparticles seem to be of more interest compared with some other conventionally used nanomaterials. ZnO NRs have some interesting properties that researchers are able to exploit. Key properties of ZnO include stability and nontoxicity for normal tissues but toxicity for carcinogenic tissues, biocompatibility, characteristic optical properties, and surface plasmon resonance (SPR) behavior. All mentioned factors provide the potential application in clinical department [17–19]. In addition, ZnO acts as an efficient nanomaterial, is basically a fluorescent compound, and can act as keystone in biological applications. The main advantages of nano-sized compounds in different biological applications are

- targeted drug delivery
- bioimaging in case of diagnostic system
- achieve enhanced permeability and retention effect

According to a literature review, the deficiency of multidrug resistance (MDR) in cytotoxicity of different biological applications was overcome by the advancement of nanotechnology. ZnO is a very a popular nanomaterial due to its tremendously notable advantages, such as stability, biosafety for normal/healthy cells, easy transport of drug vehicles across epithelia, high surface-to-volume ratio, better performance, and targeted drug delivery, ability to act as a biomarker, reduction of side effects, and interception and containment of lesions, with minimal or no loss of quality of life. In the current decade, research focused on development of versatile and minimally invasive NPs for in vivo as well as in vitro applications [20–25]. The beauty of nanoparticles drug delivery lies in that after uptake/accumulation at the targeted site, nanovectors can be triggered by an external source (ultrasound, IR, UV) to create antitumor action with the use of internal factors, e.g., pH or temperature. In the current experiment, different malignant cell lines, such as melanoma (skin cancer) and normal foreskin fibroblast cell lines, were used as biological sample/in vitro model as shown in Figs. 5.3 and 5.4. Cell culturing protocol and laser irradiation experimental steps are described as follows:

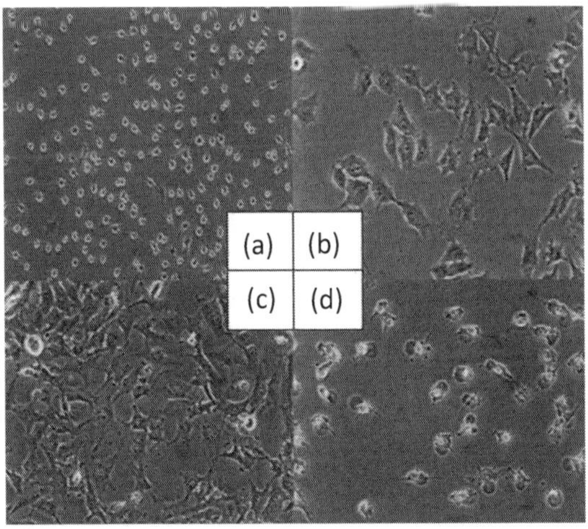

Figure 5.3 (a) Melanoma (FM55P) cell with confluence of 65–70% approximate. (b) Melanoma (FM55P) cell without serum. (c) Melanoma (FM55P) cell with serum. (d) Melanocytes (FM55P) after ALA (250 μg/mL) incubation.

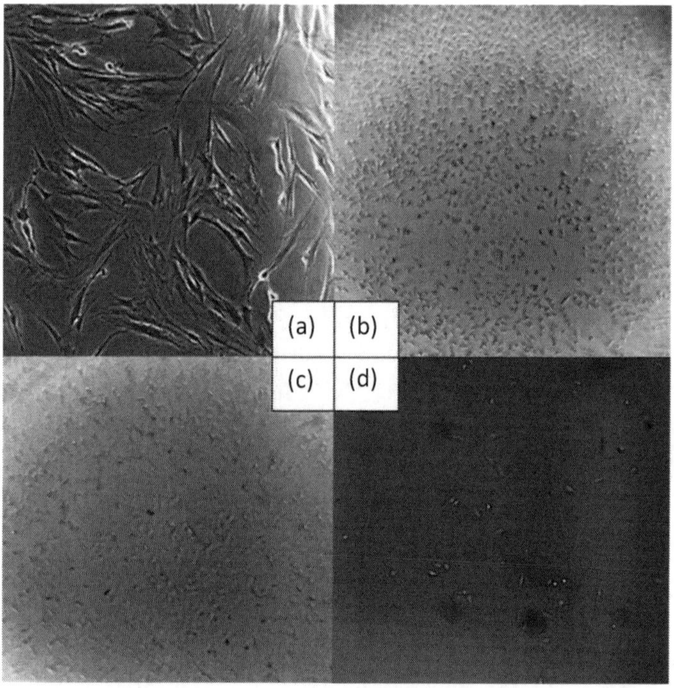

Figure 5.4 (a) Foreskin fibroblast (AG1518) after culturing growth with confluence of ≥50% approx. (c, d) Gross view (different pose) of HeLa cells in culturing media (DMEM).

5.1 Cell Culture

Melanoma (FM55P), foreskin fibroblast (AG1518), hepatocellular carcinoma (HepG2) and RD cells were seeded out in 25 cm^2 plastic tissue culture flasks (Nunc Wiesbaden Germany) in Dulbecco's modified eagle medium (DMEM) with Hanks salts, containing 10% fetal bovine serum (FBS) and 2 mM L-glutamine along with some non-essential amino acids and antibiotics (penicillin, streptomycin, and neomycin) were incubated for 24 h for proper attachment to the substratum. The aforementioned cells were maintained at 37°C in a moist environment as a sub-confluent monolayer and were routinely sub-cultured twice or thrice weekly. The cell culture with 70–80% confluence was harvested using 0.25% trypsin [26–30]. Cell confluence was assessed by inverted microscope shown in Fig. 5.5a. The cell culture process is shown in Fig. 5.5b.

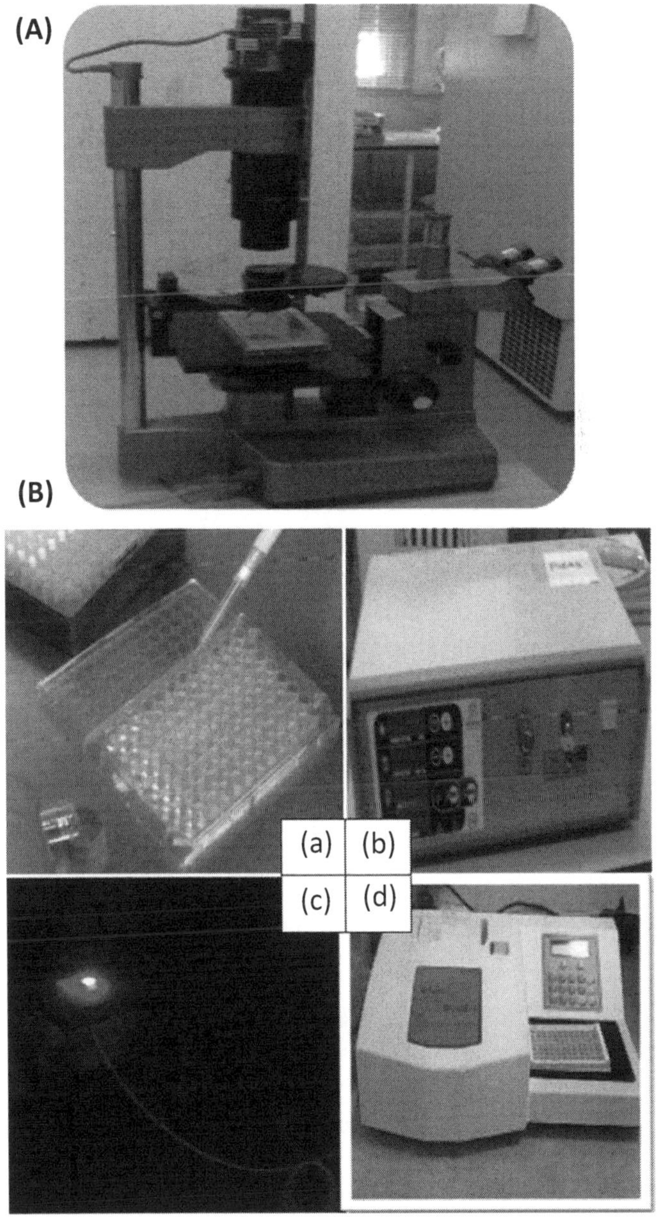

Figure 5.5 (A) Inverted microscope. (B) Laser Irradiation Portion steps: (a) Cell culturing gross view; (b) diode laser overview with 630 nm of wavelength; (c) irradiation of cells snapshot; (d) microplate reader gross view.

The rest of the experimental steps are discussed below by applying the freestanding drug delivery.

5.2 Freestanding Drug Delivery

5.2.1 Preparation of Metal Oxides Nanostructures

The hydrothermal process was used for the synthesis of all samples, i.e., ZnO NRs, MnO_2 NRs, and Fe_2O_3 NPs, and were tested for their cytotoxicity to different malignant cells, e.g., HeLa cells, HepG2 cells and RD cells melanoma and foreskin fibroblast [31–33]. The synthesis details are discussed in previously published data [34].

5.3 Stock Solution of Chemicals and Photosensitizers

The above-mentioned nanomaterials and ALA, Photofrin® and PpIX, PPDME were dissolved in phosphate buffered saline (PBS) except the PPDME, which was dispersed in dimethyl sulfoxide (DMSO), and the pH was adjusted to 7.4. This was used to prepare a solution of concentration 20 mg/mL, which was then sonicated for 40 min to obtain a complete dispersion and then it was stored in dark. For each experiment, a new aliquot was thawed immediately 3 h prior to use. The working drug solutions of varying concentrations were prepared by diluting the stock solution with PBS. Images were taken by Nikon Eclipse 400 epifluorescence microscope, and during data measurements a DAPI filter cube was used (an excitation band pass filter: 340–380 nm; dichroic mirror: 400 nm; emission band-pass filter: 435–485 nm). The images were obtained with a CCD camera and/or DU 897E.

5.4 Light Exposure

The cells were exposed to light from the top of the plate using semiconductor diode laser (LPhT-630/675-01-BIOSPEC, Russia) of 635 nm of light wavelength and a photocure AKTILITE CL 16. This is a LED-based narrow-band (630–635 nm) red light technology device, and it was illuminated up to 40 × 50 mm. The time of irradiation for the mentioned cells was kept 6.5 min with 30 J/cm^2 light dose [27, 28].

5.5 Drug Cytotoxicity and Phototoxicity Tests

For photosensitizer cytotoxicity tests, the different malignant cells were seeded at a concentration of around 1×10^5 cells/well into 96 well flat-bottomed microtiter plates and were incubated with varying concentrations of the Photofrin®, 5-ALA, PpIX and PPDME ranging 0–500 μg/mL, in a serum-free MEM at 37°C for 24 h and with nanomaterials after sonication were exposed to suggested cells for 14 h. After incubation, cellular viability was assessed by using multiscan MCC/340 microplate reader at a wavelength of 540 nm by means of neutral red assay (NRA) [27–29, 33].

5.6 Incubation of Photosensitizer and Laser Irradiation

The experiments were performed in two different steps: In the first step, the cells were exposed to different concentrations of the photosensitizers (5-ALA, PPDME and Photofrin®) for the selectivity of the optimum concentration of the photosensitizer, which was found to be 100 μg/mL. In the second the step, the cells were incubated with this optimum concentration of photosensitizers (100 μg/mL). In each step after 24 h of incubation with the photosensitizer, treated cells were then irradiated with different laser doses (0–160 J/cm^2) of a 635 nm wavelength of diode laser for assessment of the optimum light dose, which was found to be 30 J/cm^2. Six wells containing photosensitized labeled cells were exposed to this dose, while two other wells were kept as control. These experiments were repeated three times. After irradiation, the cell cultured medium (old media) was removed and the cells were washed twice in ice-cold phosphate buffer saline (PBS) with a pH of 7.4. The PBS was then replaced with a DMEM containing 10% volume FBS and the plates were placed into the incubator for 24 h. These labeled cells were processed for the assessment of cell viability as detail discussed in next few slides [26, 30–32].

5.7 Labeling of Carcinogenic Cells with MnO_2 NRs, Fe_2O_3 NPs, ZnO NRs and Their Conjugated form with 5-ALA and PPDME

Prior to the addition of the nanomaterials (as bare or conjugated with photosensitizers), the cells old medium was removed from all wells and the cells were washed twice with PBS. A new medium of 200 μL was added to the cells. Different concentrations (0.05–500 μg/mL) of MnO_2 NRs, ZnO NRs, and Fe_2O_3 NPs were added and incubated for 15 h. All working solutions were prepared in PBS and dimethyl sulfoxide (DMSO) separately. The cells were incubated in humidified air with 5% CO_2 at 37°C. The images of the labeled/photosensitized cells were taken after 15 h. The control samples were cultivated under the same conditions without the 5-ALA but with serum. Thereafter, the cells were examined and photographed using a Nikon Eclipse 400 epifluorescence microscope.

5.8 Viability Determination

The viability of the cells was measured after the exposure to different concentrations of ALA and irradiation. Cellular viability was estimated by means of neutral-red spectrophotometric assay [35]. The medium containing ALA was removed from the wells and replaced with 200 μL of fresh medium per well containing 50 μL of neutral red(conc. 50 mg/mL). The plates were then returned to the incubator for 3 h. The medium was subsequently removed, and the cultures were washed rapidly with a mixture of 40% formaldehyde and 10% $CaCl_2$ v/v (4:1). A mixture of 1% v/v acetic acid 50% v/v ethanol (1:1) was added to extract the neutral red. The plate was shaken for 60 sec and left to stand at room temperature for 15 min. The absorbance of the solubilized dye was subsequently read at 490 nm. Quantification of solubilized dye was correlated with the live cell number. Control wells were prepared parallel, and these cells were exposed to neutral red, but not to ALA. The percentage of viable cells in the cell population at each concentration of the test agent was calculated by means of this formula:

$$\%\,\text{Viability} = \frac{\text{Mean absorbance of ALA treated cells}}{\text{Mean absorbance of control cells}} \times (100)$$

5.9 Reactive Oxygen Species Measurement

For the verification of the results, intracellular ROS production was detected using the 2′,7′-dichlorofluorescein diacetate from Sigma Aldrich. The labeled cells with a range of different concentrations (0.05–500 μg/mL) were used in these experiments. Metal oxides conjugated with photosensitizers sensitized treated cells were washed gently once in Dulbecco's minimum essential medium (DMEM) and were investigated for ROS detection. Cells with 100 μM of 5 μM 2′,7′-Dichlorofluorescein diacetate were incubated in dark for 35 min at 37°C. Thereafter, the cells were exposed to UV light with a dose of 10 J/cm^2 for 2 min, and the production of the ROS in all conjugated and nonconjugated cells was measured with a POLAR Star Galaxy multifunctional microplate reader. This microscope has an excitation with a filter allowing a 485 nm excited beam. For the detection of the fluorescence, a filter of 520 nm was used during data measurements. For morphology changes, images were obtained using Nikon Eclipse 400 epifluorescence microscope with a CCD camera, and/or DU 897E. Many groups working with PDT have demonstrated in their experiments that ROS are important chemical agents for cell killing via damaging of the cell membrane, mitochondria, nucleus, or lysosomal injury [36–41].

After getting desired confluence of cell suspension, the cell was exposed to ZnO NRs (200–300 nm) bare and ZnO NRs conjugated with ALA, Photofrin with a range of working solutions (0–1000 μg/mL). After incubation of 24 h, the labeled cells were examined by applying the MTT assay and ROS detection technique under confocal and Nikon microscope. Significant losses in cell viability (≈90%) were recorded. Some investigators quoted in their publicized data mechanical stress/trauma due to size and dimensions ZnO NRs also included in the cumulative cytotoxic effects of mentioned treated cells [40].

5.10 Confocal Microscopy

When hepatocellular carcinoma (HepG2 cells) was photosensitized with 300 μg/mL working solution of bare ZnO NRs and ligands of ZnO NRs with ALA and examined by confocal microscopy, a significant amount of fluorescence was recorded, as depicted in Fig. 5.6, when HepG2 cells were incubated with 300 μg/mL of ZnO NRs. Khan et al. [42] demonstrated that ZnO—itself being a fluorescent material—shows notable bioactivity when it interacts with the metabolic activity of living tissues [43]. Meanwhile, when the above-mentioned cells were exposed to ALA complexed with ZnO, the descending order in the fluorescence was noted, because due to counterattack effect of ALA and ZnO NRs, decreasing trend in fluorescence was noted. In spite of all these results, with a loss in fluorescence, the necrotic area of the treated cell was visualized by a confocal microscope, as described in Fig. 5.7. We suggest that after labeling with ZnO NRs, remarkable amount of fluorescence was found, but no significant cell death assessed in case of HepG2 cell line, which rather is the case after the exposure of HepG2 with ALA complexed with ZnO. No dominant fluorescence was observed, but a significant number of dead cells with necrosed nucleus were visualized.

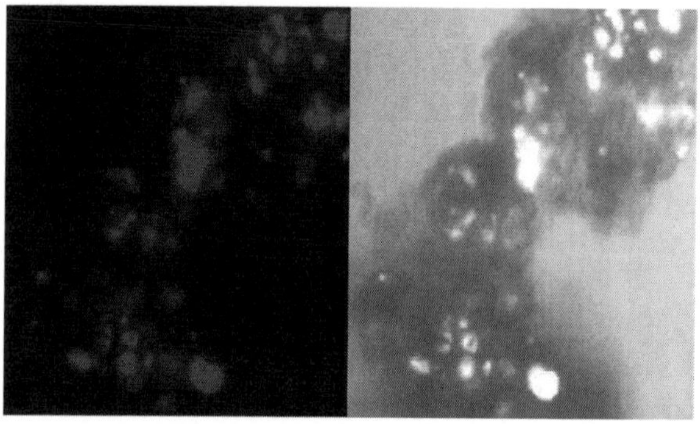

Figure 5.6 HepG2 cell labeled with ZnO NRs.

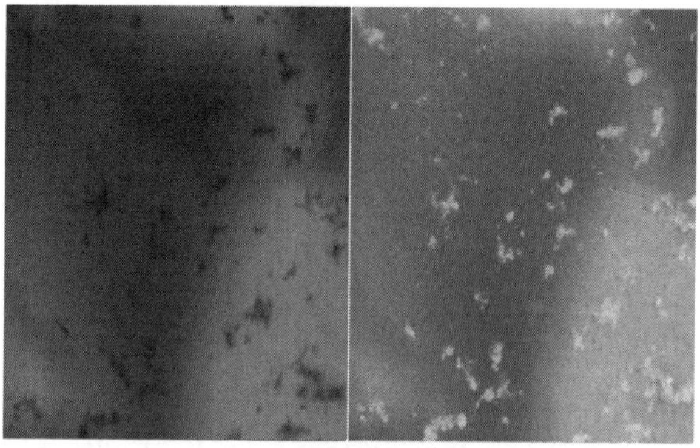

Figure 5.7 HepG2 cell labeled with ALA conjugated with ZnO NRs.

In addition, our experimental work confirms that ZnO is biocompatible and produces no toxicity in living cancer cells when it is inserted inside these cells, but whenever it is irradiated with visible or UV light, the scenario is different. Upon irradiation of ZnO that is inserted inside the cells (separately or as a complex along with photosensitizer, e.g., ALA), it affects the cytoplasm inside the cells through active or passive transport delivery system and hence enhances the fluorescence. We have previously shown that ZnO can be used as an efficient biomarker and hence can play a vital role in the field biomedical diagnosis. The present findings verify our previous results [42]. However, so far, little work has been reported on the successful applications of ZnO NPs in bio-labeling. For a few years, ZnO NPs have been used in cell imaging applications [44]. Nevertheless, a major problem in using ZnO NPs arises from their poor stability in water. ZnO nanoparticles are either hydrophobic or hydrophilic. Sometimes modifications were required to achieve hydrophilicity for hydrophobically capped ZnO nanoparticles [45]. Some other researchers experimentally demonstrated new strategies by using NPs as vehicles for the delivery of photodynamic therapy agents [46]. It was also demonstrated that a high concentration of NPs conjugated with a photosensitizer (drug) enters the lysosomes via the endocytosis, which then produces significant phototoxicity as shown when examined by trypan blue test [40, 47]. Song et al. [48] used naphthalocyanice-reconstituted low density lipoprotein (LDL) NPs for in vivo cancer imaging and treatment using HepG2 as an experimental model. They obtained excellent results by using the confocal images [48]. The results of our confocal images showed good agreement with our previously published data [42].

It has previously suggested that ZnO may be used as a green fluorescence biomarker [42, 49]. Here, the cytotoxicity of the HepG2 has also been experimentally determined by using cell culture into 96 well plates using HepG2 cell line. The cytotoxicity of different samples has been determined by using the NRA technique [26–29].

The drug delivery system holds great importance in the field of photodynamic therapy. There are different protocols for photosensitizer (drug) delivery:

(1) systematically (via injection)
(2) topically or locally (in the form of cream)
(3) orally (via juice or soft drink)

After delivery, any drug stays in the stomach for a considerable time during the process of digestion. It goes through the intestine when it is absorbed through entrohepatic circulation. At this stage, the capsule broken into the constituent drug is taken to liver for detoxification. The whole process refers to the first-pass effect [50]. After this, the drug completely dissolves in the blood stream and is then taken to the targeted area of the infected malignant tissue. A gross view of the administration of the drug is shown in Fig. 5.8; the figure shows the mechanism of drug transport, photobleaching process, and oxygenation process. In biomedical application, the size of drug/photosensitizer is the most important. However, the nano-sized particles are more preferable due to relevant size of cellular and subcellular compartments. In addition, microsized particles have limited application in the field of cancer therapy due to inadequate targeting and their being harmful to the healthy tissue via mechanical stress/trauma [43]. Nanoparticles can be used as drug delivery vehicles and biomarkers of tumors and cancer cells. Most popular examples of nanoparticles are gold, silver, zinc oxide, and titania. Due to the enhancement of permeability and retention characteristics, nanoparticles are much preferred for cancer therapy. Zinc oxide nanomaterials (ZnO NMs), due to their versatility, are involved in almost every field of life, e.g., health department/biomedical applications [51]. Data collected in multiple experiments revealed that ZnO NPs, ZnO NRs, ZnO NPS, and ZnO NWs showed a significant loss in cell viability (cell killing history) even in dark, e.g., in the absence of any UV light of 240 nm wavelength. This nullifies the previous claim that ZnO can be used as sunscreen due to its biosafety and biocompatibility characteristics. Many research experts agree that approximate 25% of cell death occurred due to size and dimension of different nanomaterials.

For cytotoxicity test, when HepG2 cells were exposed to optimal concentration of ZnO NPs of 300 μg/mL as depicted in Fig. 5.9. No significant loss in cellular viability was recorded (loss in cell viability is ≈5%). The cytotoxicity of different photosensitizers, e.g., Photofrin® and ALA along with ZnO NPs, were investigated in HepG2 cells as described in Fig. 5.10. It is clear from the figure that a considerable loss in cell viability (≈20%) was observed when HpeG2 cells were sensitized with ALA or Photofrin®. However, the case was different when HeLa cells were labeled with suitable concentration (262 μg/mL) of ZnO NRs and ZnO NRs complexed with ALA as shown in Fig. 5.11. It is clear from the Fig. 5.11 that cell killing mechanism is dependent on the drug and the cellular structure.

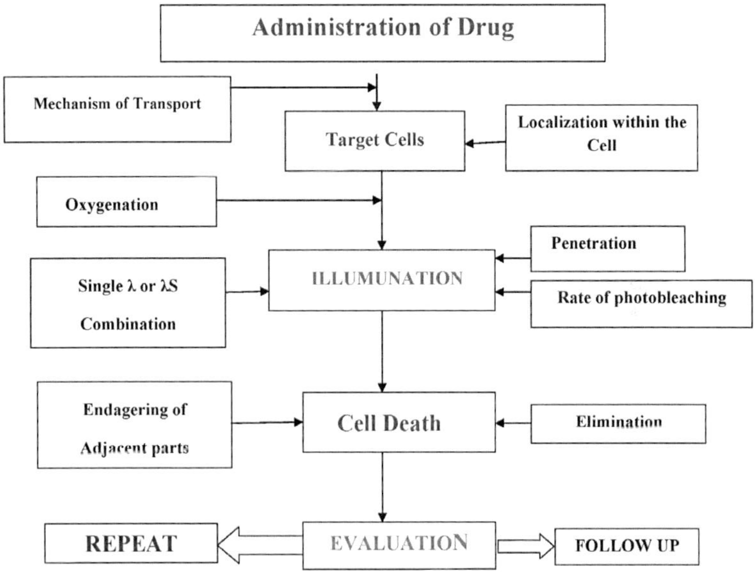

Figure 5.8 Flow chart of drug administration.

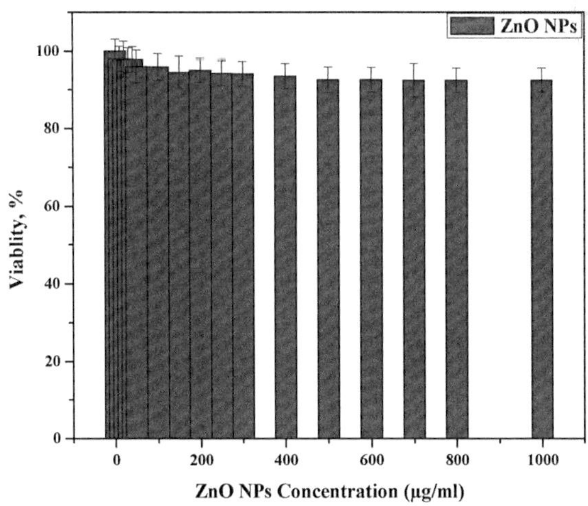

Figure 5.9 Cellular viability (%) of ZnO NPs (µg/mL) treated HepG2 cells for different concentrations. Each data point corresponds to mean $\pm\sigma$ ($n = 4$).

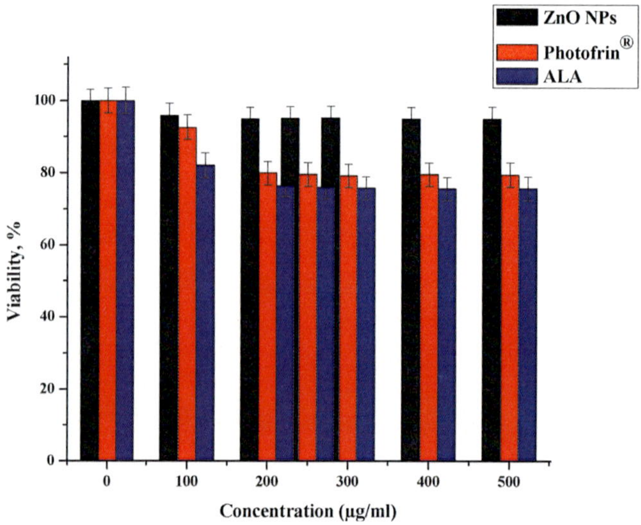

Figure 5.10 Cellular Viability (%) of (ZnO NPs (μg/mL), Photofrin® (μg/mL), ALA (μg/mL)) treated HepG2 cells for different drug doses. Each data point mean $\pm\sigma$ ($n = 4$).

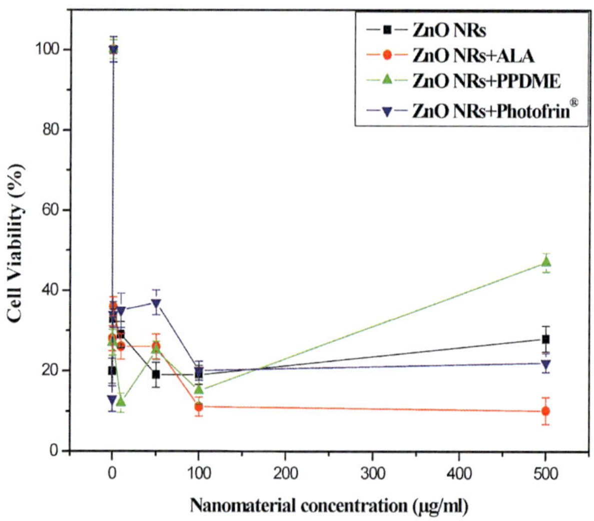

Figure 5.11 Cell viability of HeLa cells treated with different concentration of ZnO NRs and its ligands with photosensitizers.

We suggest that ZnO NRs have high potency of entering into a cell model compared with ZnO NPs or drugs. Some researcher quoted/demonstrated that ≈25% cellular loss happened due to cell rupturing membrane via the size and dimension of ZnO NRs [52]. Strangeness in current conducted experiment was observed when HeLa cells were exposed to ZnO NRs complexed with ALA. Figure 5.11 shows that after emulsion of ZnO NRs with ALA, their reinforcement to each other showed a significant loss in cell viability (≈90% loss) in case of the HeLa cell line. However, the trend of the ZnO NRs complexed with PPDME and Photofrin® is almost the same, but the prominent loss in cell viability was visualized in case of HeLa cells exposed to ZnO NRs complexed with ALA.

It was analyzed that neither the drug/photosensitizer nor the light only has the capability of induce significant cell death (≈25%) as described in the previously published data. In multiple published data, it has been already proved that a suitable dose of laser light (630 nm of red wavelength) along with an optimal concentration of photosensitizer leads to remarkable liberation of ROS/RNS, which has the potential of significant cell death via apoptosis/cell necrosis [27–29, 31, 32, 35]. The graphical pattern/template of ROS fluorescence versus drug concentration (in detail) will be plotted in future publication. From Fig. 5.12, it is clear that ZnO NRs

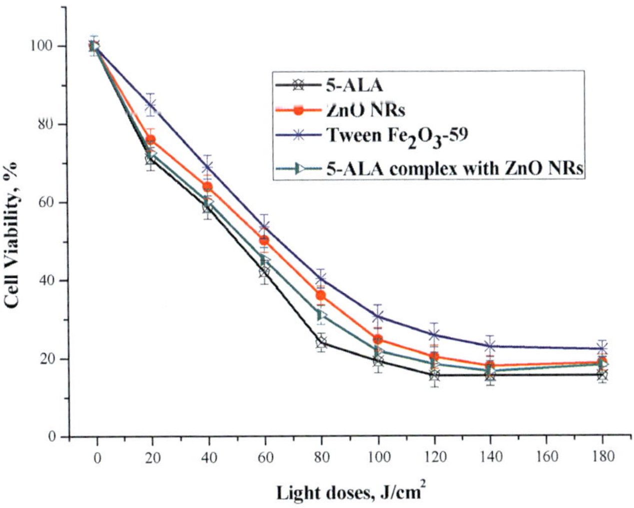

Figure 5.12 % Cellular viability of RD cells in the presence of different light doses (J/cm^2) when labeled with 5-ALA, ZnO NRs, TF-59 and 5-ALA complex with ZnO NRs. Each data point corresponds to mean $\pm\sigma$ (n = 4).

bare and conjugated with ALA show the intermediate potential of loss in cellular viability. In addition, 5-ALA individually results in the significant loss in cell viability ≈78%. An optimal dose of light (120 J/cm^2) along with 100 μg/mL of ZnO NRs indicates approximate 78% cell death toward RD cell line, which is remarkable. To conclude, ZnO NRs show a high potential of cytotoxicity toward melanoma (FM55P) as well as foreskin fibroblast (AG1518) (≈90 cell viability losses) compared with some other malignant cell line, e.g., HepG2, HeLa, and RD cell lines.

When RD cells were exposed to ZnO NRs and ZnO NRs conjugated with 5-ALA, approximate 40% loss in cell viability were found. In Fig. 5.13, it is clear that neither bare 5 ALA nor Tween Fe_2O_3-59 compounds show the significant loss in cell viability compared with ZnO NRs or ZnO NRs complex with 5-ALA toward RD cells. When muscle carcinoma were exposed to different concentration (μg/mL) of different nanomaterials, e.g., ZnO NRs, Tween Fe_2O_3-59 (TF-59) bare and conjugated with ALA (μg/mL). Comparable template of loss in cell was noticed in case of labeling of RD cells with ALA and TF-59; approximate 30% cell death factors were recorded. The case is different when the same cells were sensitized with ZnO NRs and ALA complexed with ZnO NRs; a sharp fall in cell viability was observed initially, and finally to the loss was ≈45%. We suggest that the current results (viability loss template) nullify the previous claim with the effect that ZnO is biosafe and biocompatible in dark but toxic in the presence of suitable/optimal dose of UV light (≈240 nm wavelength). Bechet et al. demonstrated that nanoparticles can act as good vehicles for the delivery of photosensitizing agent, the presence of nanoparticles show promising PDT outcomes for different malignant carcinomas [40]. Guo et al. stated that the induction of ZnO nanoparticles in the field of PDT has the excellent potential of synergistic cytotoxic effects toward malignant tissue treatments; especially, ZnO nanoparticles' contribution is to overcome MDR in the field of PDT [53]. Berardis et al. analyzed that the exposure of different sized nanoparticles toward colon carcinoma causes the induction of oxidative stress and significant cytotoxicity apoptosis and proinflammatory [54]. The published data is in good agreement with our experimental data.

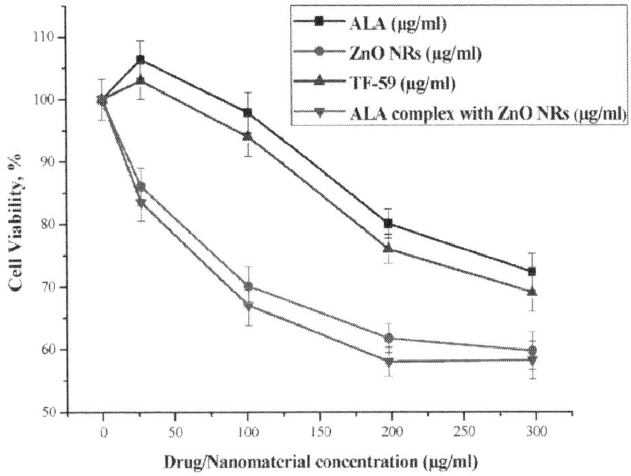

Figure 5.13 Cytotoxicity of different nano-oxides complex with photo-sensitizer in absence of light doses. Each data point corresponds to mean $\pm\sigma$ ($n = 4$).

5.11 Microinjection Drug Delivery

SEM images of bare and ZnO NR-grown Femtotip® are shown in Fig. 5.3. Microinjection drug delivery systems were introduced for the first time by our group; it is the novelty of our experiment. Chemical aqueous technique was adopted for the fine growth of ZnO NRs on the femtotip pointer [42]. This technique was developed for accurate nanoparticles drug delivery into different malignant cell lines; melanoma and foreskin fibroblasts were used as the experimental model [42]. Significant toxicity of ZnO NRs was recorded when PPDME conjugated ZnO NR- grown femtotip were used as pointer for microinjection drug delivery with and without 10 J/cm^2 of 240 nm of UV light [42].

5.12 Deposition of Nanostructures on Borosilicate Glass Capillary (Femtotip®)

Deposition of different ZnO nanostructures, such as ZnO NRs, ZnO NPs, ZnO NWs, and ZnO NFs, on the tip of Borosilicate glass capillary and its conjugation with photosensitizer Photofrin®/ALA is the focus of our present experiment (microinjection drug delivery technique). Borosilicate glass capillaries (sterile Femtotip® II with tip inner diameter of 0.5 μm, an outer diameter of 0.7 μm, and a length of 49 mm, Eppendorf, Hamburg-Germany) were used for the deposition of ZnO NPs by applying the low-temperature aqueous chemical growth technique [55, 56]. The morphology and some structural information of different nanostructures, such as ZnO NRs, ZnO NWs, ZnO NPs, and ZnO NFs, are shown in the SEM images in Figs. 5.14–5.17. By different testing techniques, such as atomic force microscopy (AFM), it was analyzed that grown ZnO NP structures vary between 200 and 600 nm in size and thus have good PDT therapeutic outcome compared with ZnO NRs [52]. The ZnO NPs grown on the capillary glass tip were functionalized by conjugating the Photofrin® layer through a manual process. The powder form of Photofrin® was purchased from Sigma Aldrich. Its molecular structure formula has been publicized in our previous data [57]. Photofrin® was dissolved in phosphate buffered saline (PBS, pH 7.4) to obtain stock solution (1000 μg/mL) [26, 30–32] and was stored in dark. The ZnO NPs-coated tips were dipped five times into the prepared working solution, which was stored in dark at 4°C. After each dip, the tip was allowed to dry at room temperature [52]. A submicrometer glass pipette covered with bare grown ZnO NPs was also used as reference PDT device to allow the separation of the contribution to fluorescence effect of the NPs and the NPs–PS conjugated tip.

Figure 5.14 SEM images of ZnO nanorods on the tip of Borosilicate glass capillary.

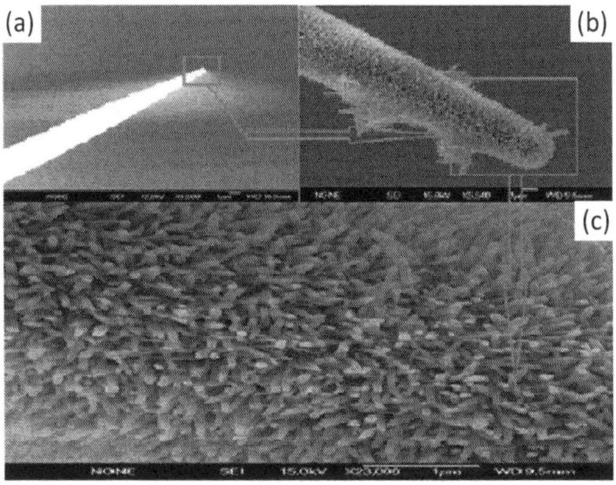

Figure 5.15 SEM image of (a) bare femtotip (b) grown ZnO-flakes (ZnO-NFs) on the tip of Borosilicate glass capillary (c) magnified image of ZnO nanoflakes.

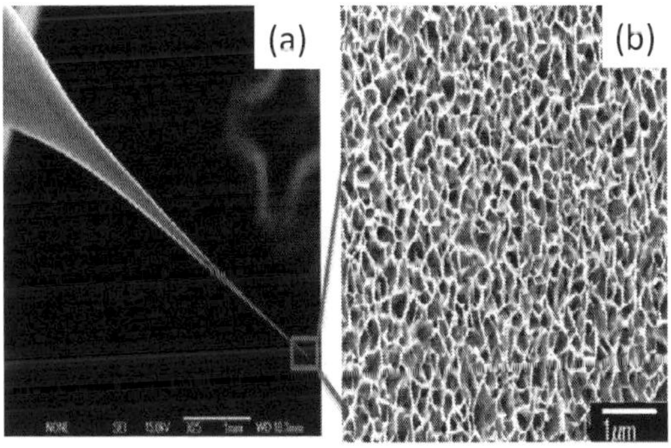

Figure 5.16 SEM image of (a) bare femtotip (b) grown ZnO nanoparticles on the tip of Borosilicate glass capillary.

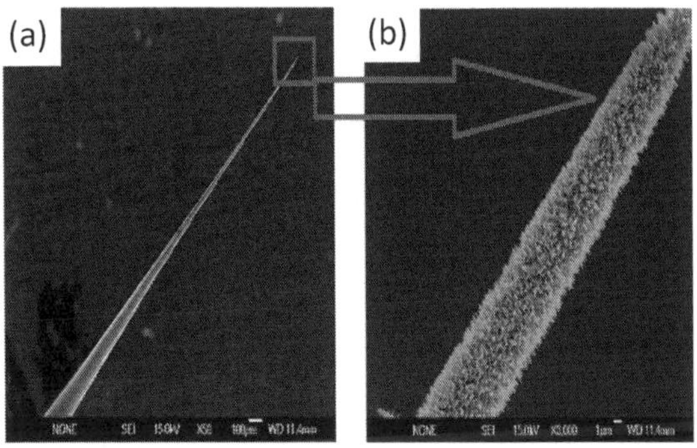

Figure 5.17 (a) SEM image of bare femtotip (b) ZnO-nanowires (ZnO-NWs) on the tip of Borosilicate glass capillary.

The fluorescence spectrum of ZnO intracellular foreskin fibroblast conjugated with femtotip excited at 240 nm of UV light is depicted in Fig. 5.19. These bare and conjugated ZnO NPs devices were used as local PDT intracellular photosensitizer delivery system for different malignant cell lines, e.g., HepG2, RD, HeLa cell treatment. During these steps, treated cells were carried under examined online Nikon microscope attached with a CCD camera.

Many researchers agree that basically nanomaterials have two categories of cytotoxicity: Initially the production of free radical ions or singlet oxygen (both consist of ROS) leads to cell death via multiple cell death mechanisms, e.g., vascular blockade, lysosomal injury or damaging effect of mitochondria. In addition, second notable parameter responsible for cell killing phenomena is the mechanical trauma/stress due to the shape, dimension, and size of different nanostructures, which is considered a drawback in the field of PDT [43]. Comparable result patterns/templates were achieved in our future publication (relevant to potency of different nanostructures toward foreskin fibroblast and melanocytes). When the femtotip need act as pointer grown ZnO NWs, inserted into after saving nucleus of foreskin fibroblast cells as different steps of microinjection insertion of femtotip are shown in Figs. 5.18(a–d). After insertion of femtotip grown ZnO NWs, treated cells were exposed to optimal dose of UV light with 20 J/cm^2; the mitochondria almost disappeared, indicating cell necrosis and the cells also seemed to shrink. Neighboring cells after 20 min UV exposure were still displaying diffused red fluorescence in the mitochondria. A similar protocol was applied for the measurement of the effectiveness of cytotoxicity via chemical excitation and mechanical stress in melanocytes too. The rest of the procedure was same except biological samples. Significant cytotoxicity was noted in case of melanoma cells after irradiation of 20 min of UV irradiation for microinjection-treated melanocytes, resulting in the disappearance of fluorescence from melanoma cells, an indication of the disappearance of mitochondria, which liberates the marvelous negative effect for the production of ATP/ADP into the cells. Mitochondrial disappearance from cells can lead to apoptosis. Some researchers have claimed that ZnO is biocompatible and biosafe [58]. However, in this experiment, our results nullify the previous statement regarding biosafety and biocompatibility of different ZnO NRs materials with concluding remarks of notable cytotoxicity of ZnO NRs for fibroblast as well as melanoma cells under irradiation of suitable UV light. Figure 5.18a shows the SEM image of bare femtotip before the beginning of any step of the experiment. In addition, Fig. 5.18b depicts the excellent description of ZnO NR–grown femtotip in form of the SEM image model. When the ZnO NRs–grown femtotip were inserted into foreskin fibroblasts (AG1518) as shown in Fig. 5.18c, but beauty of this conducted experiment is the saving of nucleus from any trauma/mechanical stress and enhancement of photochemical reactions under irradiation of UV light of 240 nm wavelength. The consistency in the ROS fluorescence (as shown in Fig. 5.18d) was visualized as increasing the UV light dose of 10 J/cm^2 to 20 J/cm^2 with sensitized coating of 200–300 nm of ZnO NRs.

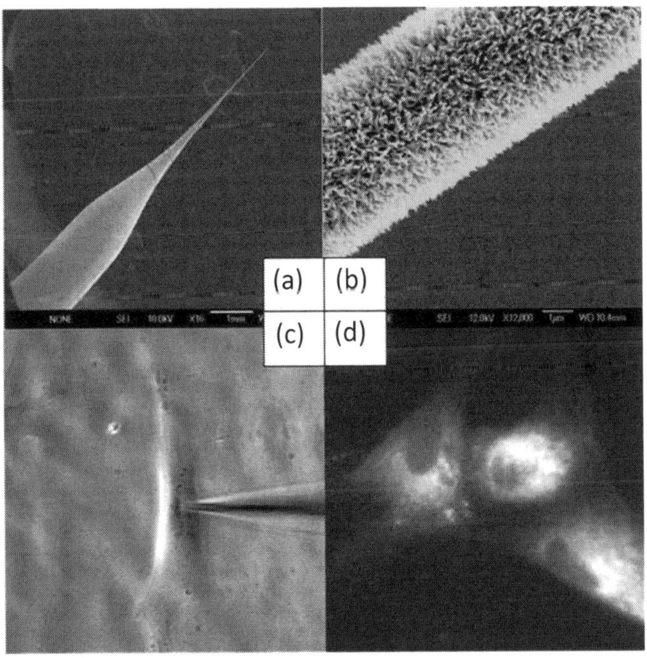

Figure 5.18 SEM image of (a) bare femtotip (b) grown ZnO-wires (ZnO-NWs) on the tip of Borosilicate glass capillary (c) Insertion of femtotip in fibroblast cells (d) liberation of ROS after 10 J/cm^2 of UV irradiation.

Cytotoxic effects with ZnO NR–grown femtotip under UV light were studied in the foreskin fibroblast cell line. Such study has not been conducted previously. By fluorescence spectroscopy, ZnO

NRs displays four dominant peaks: One falls in the UV region at 501 and 510 nm and the rest of the peaks fall at 560 and 575 nm, as shown in Fig. 5.19, when spectrofluorimetery of ZnO NR-grown femtotip intracellular foreskin fibroblasts were entertained.

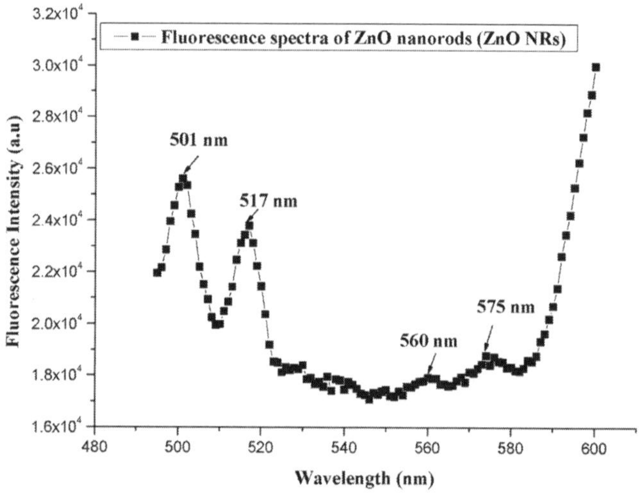

Figure 5.19 Fluorescence spectrum of ZnO NRs intracellular foreskin fibroblast (AG1518).

Figure 5.20 shows a schematic diagram the PDT mechanism that represents the good algorithm of photodynamic reactions excited with 240 nm of UV light, also part of the current experimental setup. After the insertion of ZnO NR-grown femtotip insertion are shown in snapshot, but when excited with suitable intensity of UV light, it enhances the excitation of the foreskin fibroblast intracellular ZnO NRs resulted with the release of ROS and leads to cell necrosis within a few minutes. The cell killing mechanism consists of basically six steps shown in Fig. 5.20. In step 1, intracellular ZnO NRs were excited via 240 nm of UV light (light dose of 10 J/cm^2). After the excitation of intracellular ZnO NRs, white light with broad wavelength obtained as shown in step 2 and red light with a wavelength of 630 nm were capable of exciting Photofrin®, which was conjugated with ZnO NRs before insertion, as depicted in step 3. As Photofrin® conjugated with ZnO NRs is excited by indirect red light (emitted by ZnO NRs after UV excitation), it is therefore consequently the emitted light absorbed by Photofrin (630 nm) [59]. In turn, ROS is released, which is responsible for damaging the mitochondria as depicted in steps 4 and 5. In fact, ultimately the liberation of ROS acts as appetizer for the cell killing mechanism via mitochondria damaging effect as shown in step 6. Many research experts agree that the optimal dose of light (UV of 240 nm, 20 J/cm^2) along with a suitable ZnO/drug concentration proves to be good source of active species of oxygen, which leads to the mitochondria-damaging mechanism/lysosome-damaging effect responsible for the cell killing [60].

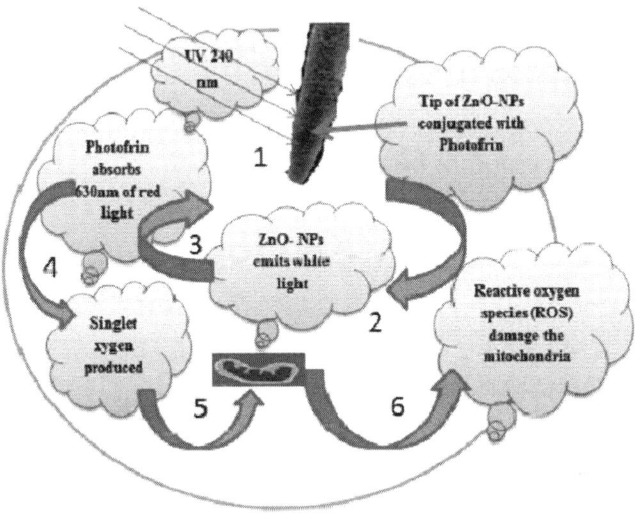

Figure 5.20 Schematic of PDT mechanism.

Photodynamic effects of different ZnO nanostructures are analyzed in neoplastic cells of melanocytes (A375P), normal human foreskin fibroblast cells (AG01518), RD and hepatocellular carcinoma freestanding and intracellular insertion of ZnO nanorods tip coated with drug (photosensitizer)/exposure to drug (PS) by analysis of liberation of ROS detection, mitochondrial staining procedure, microscopic study and cell viability. ZnO nanorods can be used for the tumor-selective delivery of chemotherapeutic agents. Furthermore, it is examined that there is synergetic effect of ZnO nanorods with Photofrin® under UV irradiation, indicating that ZnO NRs are potent photosensitizers. ZnO nanorods alone or conjugated with ALA possess a similar template of effectiveness in foreskin fibroblast cells. This observation almost nullifies the efficacy of ALA in both normal as well as cancerous cells, suggesting that anticancer photosensitizers (ALA and PPDME) have the least penetration into the normal cells—a selective property of photosensitizers. However, ZnO nanorods, not the classical anticancer drugs, offer maximum penetration into the normal cells and thus pose toxic effects for normal tissues, which negates the previous claim that they are biosafe and biocompatible. To conclude, Photofrin® has remarkable antitumor effectivity on melanoma cells. Therefore, we believe that Photofrin® can be an appealing candidate for the treatment of melanocytes as well as hepatocellular carcinoma.

Acknowledgements

We acknowledge Prof. Karin Olinger and Kishwar for technical assistance in collecting data for this work.

References

1. M. Riaz et al., *Nanotechnology*, **19**, 415708 (2008).
2. M. Willander, et al., *Spie Proc.*, **6486**, 648614 (2007).
3. S. M. Al-Hilli, M. Willander, A. Öst, and P. Strålfors, *J. Appl. Phys.*, **102**, 084304 (2007).
4. X.D. Bai, P.X. Gao, Z.L. Wang, and E. G. Wang, *Appl. Phys. Lett.*, **82**, 4806 (2003).
5. Z. L. Wang and J. H. Song, *Science*, **312**, 242 (2006).
6. Y. Qin, X. D. Wang, and Z. L. Wang, *Nature*, **451**, 809 (2008).
7. Q. X. Zhao, et al., *Phys. Scr.*, **126**, 131 (2006).
8. N. H. Alvi, M. Riaz, G. Tzamalis, O. Nur, and M. Willander, *Semicond. Sci. Technol.*, **25**, 065004 (2010).
9. N. H. Alvi, K. Hassan, O. Nur, and M. Willander, *Nanoscale Res. Lett.*, **6**, 130 (2011).
10. J. R. Sadaf, M. Q. Israr, S. Kishwar, O. Nur, and M. Willander, *Nanoscale Res. Lett.*, **5**, 957 (2010).
11. J. P. Santos and J. A. de Agapito, *Thin Solid Films*, **338**, 276 (1999).
12. N. H. Alvi, S. M. Usman Ali, S. Hussain, O. Nur, and M. Willander, *Scripta Mater.*, **64**, 697 (2011).
13. N. Hamada, S. Sawada, and A. Oshiyama, *Phys. Rev. Lett.*, **68**, 1579 (1992).
14. W. Li, et al., *J. Photochem. Photobiol.*, **60**, 79 (2001).

15. C. S. Xu and A. W. N. Leung, *Laser Phys. Lett.*, **7**, 68 (2001).
16. S. M. Hussain, K. L. Hess, J. M. Gearhart, K. T. Geiss, and J. J. Schlager, *Toxicol. in Vitro*, **19**, 975 (2005).
17. M. C. Daniel and D. Astruc, *Chem. Rev.*, **104**, 293 (2004).
18. I. H. El-Syed, X. Huang, and M. A. El-Sayed, *Nano Lett.*, **5**, 829 (2005).
19. K. V. Sarathy, G. Raina, R. T. Yadav, G. U. Kulkarni, and C. N. R. Rao, *J. Phys. Chem. B.*, **101**, 9876 (1997).
20. Y. Nishioka and H. Yoshino, *Adv. Drug. Deliv. Rev.*, **47**, 55 (2001).
21. J. E. Schnitzer, *N. Engl. J. Med.*, **339**, 472 (1998).
22. S. Kim, et al., *Nat. Biotech.*, **22**, 93 (2004).
23. M. Ferrari, *Nat. Rev. Cancer*, **5**, 161 (2005).
24. S. A. Wickline and G. M. Lanza, *Circulation*, **107**, 1092 (2003).
25. W. Zhao, *Nanomedicine*, **1**, 238 (2005).
26. M. Fakhar-e-Alam, et al., *Laser Phys.*, **21**, 982 (2011).
27. M. Atif, et al., *Laser Phys. Lett.*, **7**, 757 (2010).
28. M. Atif, M. Fakhar-e-Alam, S. S. Z. Zaidi, and R. Suleman, *Laser Phys.*, **21**, 1135 (2011).
29. M. Atif, et al., *Laser Phys.*, **21**, 1235 (2011).
30. M. Fakhar-e-Alam, M. Atif, T. Rehman, H. Sadia, and S. Firdous, *Laser Phys.*, **21**, 1428 (2011).
31. M. Fakhar-e-Alam, et al., *Laser Phys.*, **21** (2011); DOI: 10.1134/S1054660X1119011X.
32. M. Fakhar-e-Alam, et al., *Laser Phys.* (2011); DOI: 10.1134/S1054660X11210067.
33. M. Atif, et al., *Laser Phys.* **21**, 1 (2011), in press.
34. Y. Khan, S. K. Durani, M. Mehmood, and M. R. Khan, *Mater. Lett.* (2011); DOI: 10.1557/jmr.2011.138.
35. M. Atif, et al., *Laser Phys. Lett.*, **6**, 886 (2009).
36. A. M. Derfus, W. C. W. Chan, and S. N. Bhatia, *Nano Lett.*, **4**, 11 (2004).
37. H. Xiong, et al., *Adv. Funct. Mater.*, **15**, 1751 (2005).
38. X. Tang, E. S. G. Choo, J. D. L. Li, and J. Xue, *Langmuir*, **25**, 5271 (2009).
39. M. W. Yvonne, W. Janssen, S. Matalon, and B. T. Mossman, *AJP-Lung Cell. Mol. Physiol.*, **273**, L789 (1997).
40. D. Bechet, et al., *Cell Press*, **26**, 612 (2008).
41. P. Alivisatos, *Nat. Biotechnol.*, **22**, 47 (2004).

42. Y. Khan, S. K. Durrani, M. Mehmood, M. R. Khan, and S. Firdous, *Appl. Surf. Sci.*, **257**, 1756 (2010).
43. J. Li, et al., *Nanoscale Res. Lett.*, **5**, 1063 (2010).
44. D. Vollath, D. V. Szabo, and S. J. Schlabach, *Nanopart. Res.*, **6**, 181 (2004).
45. J. Jin, G. K. Soon, H. Y. Jung, and H. Taegwan, *Adv. Mater.*, **17**, 1873 (2005).
46. J. Schwiertz, A. Wiehe, S. Grafe, B. Gitter, and M. Epple, *Biomaterials*, **30**, 3324 (2009).
47. K. Chen, et al., *J. Photochem. Photobiol. B: Biol.*, **96**, 66 (2009).
48. X. L. Xu, et al., *Growth*, **223**, 201 (2001).
49. Y. L. Wu, et al., *Nanotechnology*, **19**, 1 (2008).
50. L. S. A. B. C. Yu, D. Pharmaceutical Technology: Biopharmaceutics, ed. Dekker Encyclopedias: Taylor and Francis Books, NY–USA (2002).
51. J. Chen, et al., *J. Exp. Nanosci.*, **3**, 321 (2008).
52. S. Kishwar, M. A. Hameed, O. Nur, M. Willander, and P. O. Larsson, *Nanoscale Res. Lett.*, **5**, 1669 (2010).
53. D. Guo, et al., *J. Photochem. Photobiol. B Biol.*, **93**, 119 (2008).
54. B. D. Berardis, et al., *Toxicol. Appl. Pharmacol.*, (2010); doi.org/10.1016/j.taap.2010.04.012.
55. A. Fulati, et al., *Sens. Actuators B*, **150**, 673 (2010).
56. N. H. Alvi, S. M. U. Ali, S. Hussain, O. Nur, and M. Willander, *Scripta Mater.*, **64**, 697 (2010).
57. M. Fakhar-e-Alam, et al., *Radiochim. Acta*, **98**, 813 (2010).
58. P. W. Sadik, S. J. Pearton, D. P. Norton, E. Lambers, and F. Ren, *J. Appl. Phys.*, **101**, 104514 (2007).
59. P. Kubat, Z. Zelinger, and M. Jirsa, *Radiat. Res.*, **148**, 382 (1997).
60. H. Kolaroval, et al., *Gen. Physiol. Biophys.*, **27**, 101 (2008).

Chapter 6

Zinc Oxide Nanostructures: Synthesis, Characterization, and Device Applications on Nonconventional Substrates

Magnus Willander, Omer Nur, Kamran ul Hasan, Gul Amin, and Muhammad Yousuf Soomro

Linköping University, ITN, Norrköping, S-60174, Sweden

magwi@itn.liu.se

This chapter presents an overview of the growth behavior of zinc oxide (ZnO) nanostructures on the nonconventional substrates such as paper, plastics, and polymer, with special attention to recent developments in ZnO nanostructures growth on paper substrates and their utilization for different applications. Various methods are available in the literature for the synthesis of ZnO nanostructures, the selection of which is often dependent on the desired properties and final application. Most of the synthesis methods demand high temperatures, which restricts the use of solid substrates and hence lack the applications where there is requirement of flexible and disposable substrates. In this chapter we will focus on the low-

Zinc Oxide Nanostructures: Advances and Applications
Edited by Magnus Willander

ISBN 978-981-4411-33-2 (Hardcover), 978-981-4411-34-9 (eBook)
www.panstanford.com

temperature (<90°C) solution growth method, which is highly compatible with paper substrates and does not require severe growth conditions. In addition, low-cost and large-scale mode of flexible electronics with reasonably high performance is required for many applications. In modern devices, semiconductors and metallic contact are the prime requirements. We demonstrate a very easy method to fabricate ZnO UV sensor, made on common pencil-drawn circuit over a paper. ZnO nano-crystals were extracted in a high throughput via a simple and green route. This sensor is well capable of detecting UV light and demonstrates features comparable to the sensors made with complex and expensive techniques. Other devices based on ZnO nanostructures grown on paper, by either direct growth or using scalable printing technique, such as light-emitting diodes (LEDs) and energy-harvesting components, are to be demonstrated.

6.1 Introduction

Zinc oxide (ZnO) is an important semiconductor material with its wide variety of potential applications such as LED [1], biosensor [2], pressure sensors [3], ultraviolet (UV) photo-detector [4], solar cell [5], piezoelectricity, and power generation [6]. ZnO has direct bandgap of 3.37 eV, large excitons binding energy of 60 meV, high thermal/chemical stabilities, piezoelectricity and the option of wet chemical etching, etc. [7, 8]. ZnO is now seen as an alternative material to the nitride semiconductors. ZnO possess a rich family of nanostructures such as nanorods, nanowires, nanobelts, nanoparticles, nanotips, and nanotubes [7–9]. P-type doping of ZnO is still a problem that weakens the prospects of a ZnO p–n homojunction device [10]. Alternatively, ZnO is naturally n-doped and does not need external dopants. A unipolar device made of ZnO seems to be a very a reasonable approach.

Most UV sensors have relied on traditional metallic contacts to the ZnO. Typically, the requirement of low cost has been realized by the use of lithographic techniques to produce metal electrodes. Screen-printed (thick-film) graphite electrodes have also been proposed for facilitating cost-effective fabrication process. However, it needs to be noted that some of these electrodes are only disposable in the economic sense, as relevant occupational health and environmental issues may still be in jeopardy. New, easy to make and renewable electrodes are needed for accelerating the realization of large-scale device manufacturing.

The core in pencil is made of graphite, where graphene sheets are connected together by weak van der Waals attractions with inter-laminar shear strength of only 26 MPa [11, 12]. While writing with a graphite pencil, we are actually all drawing out numerous graphenes from the graphite rod of the pencil [13], illustrated in Fig. 6.1. We use very simple, pencil-drawn interdigitated electrodes and screen-printed ZnO nanocrystals (NCs) to achieve a UV sensor on paper. This graphitic circuitry has many other advantages over conventional metal electrodes; for instance, it is strong, light, and corrosion- and heat resistant, besides being cheap and easy to fabricate.

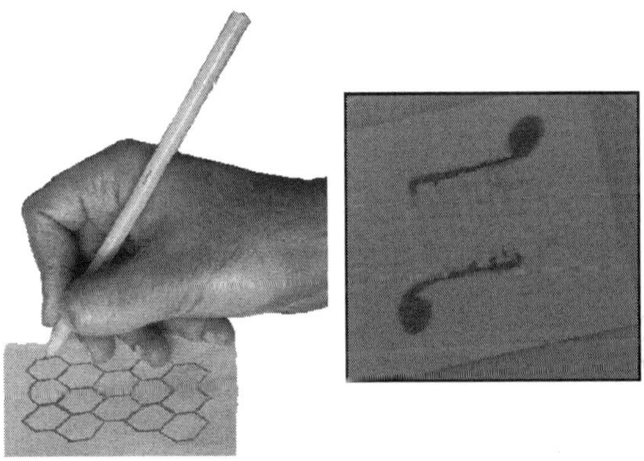

Figure 6.1 Illustration of graphene-pencil analogy (left), the picture of the pencil-drawn ZnO device.

ZnO NCs were extracted from the solution bath used for the growth of ZnO nanowires (NWs). The NWs are typically grown from solution-based technique to fabricate devices such as LEDs [1, 14, 15] and biosensors [2, 16] by uploading the samples upside down in the growth solution (not shown). We collected ZnO NC from the chemical bath after the growth of NW on some samples and dried these in powder form, consequently making a printable and semiconducting ZnO NC paste by employing a commercial binder. Illustration of graphene-pencil analogy (left), the picture of the pencil-drawn ZnO device is shown in Fig. 6.1. The morphology of the ZnO nanorods was investigated using scanning electron microscope (SEM). Figure 6.2a shows the SEM image of the bare paper that is used as a substrate, the surface of which could be seen to be very rough. Generally, paper is synthesized from cotton; major components of paper are cellulose and $CaCO_3$. Figure 6.2b shows the SEM image of the ZnO NRs grown on a cyclotene coated paper substrate and the length of the NRs were found to be around 2 μm. The growth of ZnO NRs also occurs on a bare paper substrate and the lengths of the ZnO NRs were about 1–1.2 μm while the diameter was approximately 100–120 nm. However, unlike bare paper substrate the ZnO NRs on cyclotene treated substrate was uniform and covered the substrate completely. This also confirms that ZnO NRs can be grown on any kind of substrate independent of the morphology of the underlying surface. The ZnO NRs will have different size distribution depending on the surface used for the growth.

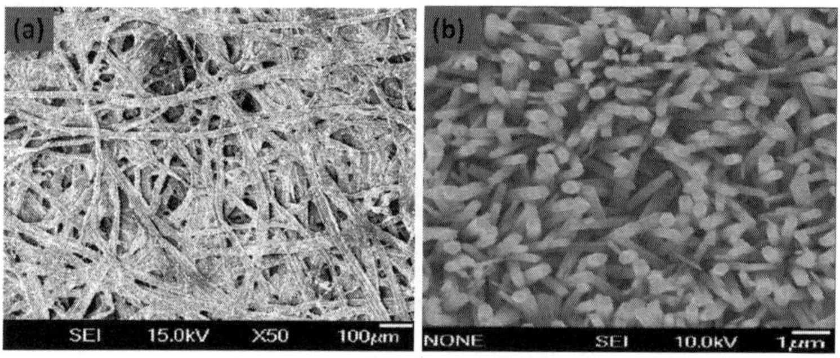

Figure 6.2 (a) SEM image of the bare paper that is used as a substrate, (b) SEM image of ZnO NRs grown on a bare paper substrate.

6.2 Experimental

Zinc nitrate hexahydrate ($Zn\ (NO_3)_2 \cdot 6H_2O$) was dissolved in 100 mL deionized (DI) water to attain 0.1 M concentration, and 0.1 M of hexamethylenetetramine (HMT, $C_6H_{12}N_4$) in 100 ml DI water [14, 17, 18]. The solution was kept at 90°C for 3 h and ZnO NCs were extracted and dried in powder form from it. XG383 Plastijet XG Varnish was used as a binder to make a ZnO NC ink for screen printing. Different grade pencils were tested for resistivity and standard 4B pencil was chosen with a line resistance of ~3 MΩ. A 4B pencil typically has 4:1 graphite to clay binder ratio. Our

ZnO NC film has a horizontal sheet resistance of ~6–7 MΩ; hence, there is no need for very high conductive sophisticated electrodes for the sensor fabrication. ZnO NC paste was screen printed onto the pencil-drawn interdigitated circuit and characterized by scanning electron microscopy (SEM), X-ray diffraction (XRD) and transmission electron microscopy (TEM). ZnO NCs are visible in the SEM image of the screen-printed film shown in Fig. 6.3a. XRD and transmission electron microscopy (TEM) verify good crystalline quality of the ZnO (Figs. 6.3b–d). The atomic force microscope (AFM) image of the pencil-drawn lines is shown in Fig. 6.3e, and the topography of the ZnO printed film in depicted in Fig. 6.3f.

Figure 6.3 (a) SEM image of the printed ZnO NCs, (b) XRD of the ZnO NCs printed on the paper, (c) Low-resolution TEM, and (d) HRTEM image of a ZnO NCs showing good crystal quality, (e) AFM image of the pencil surface, and Optical profilometer image of the ZnO printed film topography.

6.3 A Novel Low-Temperature Direct Growth of ZnO Nanorods on Paper Substrates

There is a great demand in every field such as optoelectronics, sensing, and medical to achieve soft portable electronic devices that are low-cost, smart, and multifunctional. Such devices must be reproducible on low-cost flexible and foldable substrates, e.g., plastic, paper, etc., to create smart chip cards, roll-up screens, and even disposable displays. Also, the active materials of the device must be processable by solutions, printing and the production process must be high volume technique, together with a low environmental impact. Research is actively focused on using nonconventional substrates for basic electronics components, which make excellent alternative substrates with exceptional technological attributes and commercial perspectives for the many substrates available. On approach to fulfilling these demands is to fabricate the devices on regular paper as an alternative to hard substrate, because it is one of the most abundant material and a recyclable resource. Despite these bright forecasts, little attention has been paid to disposable electronic devices or using paper as a physical support. This is mainly due to the difficulty in growing semiconductors on paper. Various electronic devices based on purely organic materials have been already realized such as paper batteries [19], electronic paper displays (EPDs) [20], thin-film transistors [21–23], photodiodes [24], and biomedical applications, [25] on such a potential substrate. Various methods such as chemical, thermal, spin coating, spray pyrolysis, and pulsed laser deposition have been used for ZnO thin-film and nanostructures synthesis but they are limited to solid supports such as metal, metal oxides, glass, and other thermally stable substrates. By being able to grow these nanostructures directly onto the substrates, scientists hope to get rid of the often energy-consuming and expensive transfer processes. This has already been demonstrated on a large scale on metal substrates. In these applications, researchers transfer thin films, nanoparticles, or other nanostructures onto the desired substrate via various processes. On the other hand, direct growth of 1D nanostructure on such paper substrates is rare and there exist only a few reports on paper substrates despite its known technological applications.

ZnO is a very promising II–VI direct wide bandgap (3.4 eV) material and provides scope for different electronics and opto-electronics applications [21, 22, 26]. Among these applications is the development of intrinsic white LEDs due to the existence of large number of radiative deep levels in ZnO material [23]. On the other hand, during the past few years, ZnO nanostructures and especially nanorods (NRs) have attracted extensive research attention because of their unique properties and potential applications, combined with the possibility of low-cost, large-area deposition, and low-temperature growth on a variety of substrates [27]. ZnO nanostructures can be grown either by low (<90°C) or high-temperature methods and both have some advantages and disadvantages. High-temperature growth methods need sophisticated equipment, which are costly. In contrast, the low-temperature growth methods have a cheaper production cost and can be used for covering large area and on soft substrates. Researchers have grown ZnO nanostructures at low temperature on a wide variety of hard and brittle substrates such as 4H-SiC [28], Si [24], GaN [29], and on flexible plastic substrates [27]. These substrates cannot be used for the applications in which there is a requirement for a cheap foldable or disposable nature.

If this problem can be addressed by employing ZnO NRs grown on a paper or paper-like substrates then low-cost, biocompatible, and disposable electronic devices such as smart cards, wearable displays, and inorganic/organic LEDs can be realized. We have fabricated ZnO nanostructures on different nonconventional substrates and the structures was utilized for ZnO nanorod-polymer LED by the hydrothermal method on paper substrate at very low temperature (50°C). Recently researchers in Taiwan have addressed the issues that they can directly grow vertically aligned, highly crystalline, and defect-free single-crystalline zinc oxide nanorods and nanoneedles on paper [30]. Although the report presented an inorganic/organic hybrid junction diode, the authors have reported for the application ultra-violet (UV) photo detector. Similarly there are many researched who reported different application utilizing ZnO nanostructures on paper substrates [31, 32].

6.4 Results

Photoconductive response is a key parameter for a photodetector. Response of our paper devices was measured using a conventional 365 nm UV lamp (Fig. 6.4a). The real time ON/OFF switching was measured by turning the UV lamp on and off inside a dark room. The measured photocurrent under the unfocused UV source shows a fast rise upon exposure to UV light and a relatively slower fall once the UV source is turned off as seen in Fig. 6.4b. Photocurrent pulse shows good stability and reversibility. Under dark condition, oxygen molecules are adsorbed on the NC surface and capture free electrons from the n-type ZnO, making negatively charged O_2 ions at the surface. This creates a low conductivity depletion layer near the nanowire surface.

$$O_2\,(g) + e^- \rightarrow O_2^-\,(ad)$$

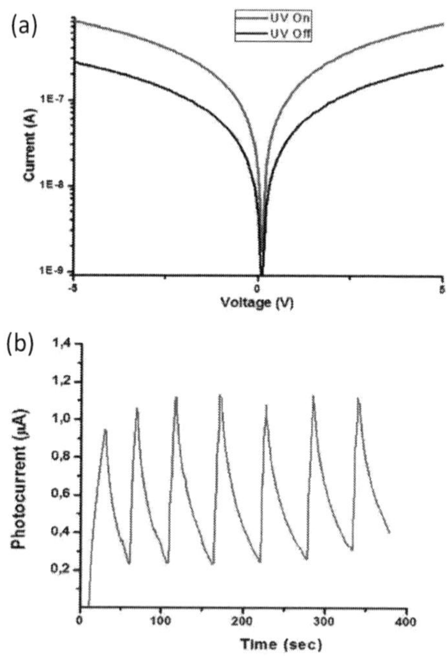

Figure 6.4 (a) *I–V* curves of the flexible pencil-drawn UV sensor with and without UV illumination, (b) Time-dependent photo response of the flexible UV sensor at a bias voltage of 1 V under UV irradiance from an unfocused UV lamp irradiating at 365 nm.

When the UV exposure is made, electron–hole pairs are photogenerated and holes are trapped at the surface by the oxygen ions via surface electron–hole recombination. However, unpaired electrons are left behind, which add to the photocurrent [4, 33, 34].

$$h^{+} + O_2^{-}(ad) \rightarrow O_2(g)$$

When the UV illumination is switched on or off, the oxygen is desorbed or re-adsorbed in the interfacial region in the premises of the metal contact in Schottky diodes, which reduces the barrier height that makes it act as a valve. However, this phenomenon takes place throughout the bulk of the material with no valve-like control, for a sensor with an absence of such a barrier. Thus, in our device a lot of unpaired electrons are generated a across the ZnO film, which results in a rapid rise in the photocurrent. Once the UV exposure is turned off, it takes some time to drain out all those photogenerated electrons. This is the reason for the faster current rise time and much slower photocurrent decay, but it is still fast enough in comparison with recent reports on ZnO based flexible UV sensors [35].

To fabricate ZnO nanorod-polymer hybrid WLEDs, we took ordinary clean room paper and sliced it into small pieces (70 mm × 70 mm). Paper has an insulating and rough surface and it can degrade in water and other nutrient solutions. In order to overcome these difficulties we have modified the surface of the paper substrate by spin coating a cyclotene layer (BCB 4022-35 Dow Chemicals), followed by vacuum baking at 110°C for 1 h. The cyclotene barrier layer, not only protects the surface of the paper from the nutrient solution and water during processing, but it also reduces the surface roughness. After that we have filtered the poly (3,4-ethylenedioxythiophene) poly(styrenesulfonate) (PEDOT:PSS) solution with 0.45 μm pore size filter, which was then spin coated on the entire substrate at a speed of 2000 rpm, followed by baking at 70°C for 2 h. We covered some area to be used for bottom contacts and then we spin coated the poly (9,9-dioctylfluorene), PFO (American dye source ADS 129 BE) conducting blue emitting polymer layer using a speed of 2000 rpm, followed by baking at 70°C for 10–15 min. In order to achieve the ZnO NRs growth we first spun-coated (four times) the ZnO nucleation layer, prepared following the method developed by Pacholski et al. [25], at a speed of 3000 rpm for 30 sec. The growth of the ZnO NRs was conducted using the hydrothermal method at 50°C temperature. The growth solution used in this method contains 150 mM of zinc nitrate hexahydrate ($Zn(NO_3)_2 \cdot 6H_2O$) and 100 mM of hexamethylene-tetramine (HMT, $C_6H_{12}N_4$). We took the pre-coated paper substrate and loaded them into a glass beaker containing the nutrient solution; these were then kept for several hours in an ordinary laboratory oven at a temperature of 50°C [24]. Finally, the samples were collected, cleaned with deionized water several times, and dried in air. In order to process the samples we spun-coated a polymethyl methacrylate (PMMA) layer, to electrically isolate neighboring nanorods from each other, at a speed of 3000 rpm for 30 sec and then baked at 90°C for 1 min. Reactive ion etching (RIE) was used to etch the PMMA layer to expose the tips of the NRs [30]. We then evaporated 50 nm aluminum (Al) top contacts and the bottom contact to the PEDOT: PSS was achieved by applying silver (Ag) paste.

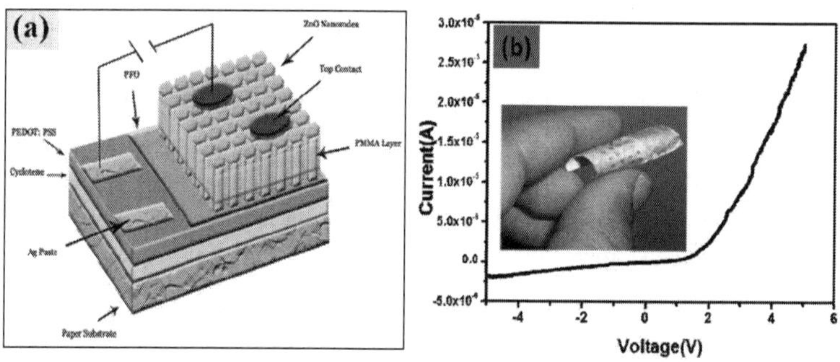

Figure 6.5 (a) Schematic illustration of ZnO/Polymer hybrid LEDs. (b) Typical *I–V* characteristics of the fabricated device on a paper and the inset shows the digital photograph of the fabricated devices on a paper substrate containing 24 devices and rolled from one side.

Figure 6.5a shows a schematic diagram of the device, while Fig. 6.5b shows the current–voltage (*I–V*) characteristics of the fabricated ZnO-polymer LED consisting of Ag/PEDOT-PSS/PFO/ZnO NRs/Al. The *I–V* characteristics indicate clearly that the diode behavior is rectifying. The rectification ratio was 10 at 4 V and the reverse leakage current was found to be 1.8 μA at −5 V. The semi-log *I–V* plot (not shown) exhibits ohmic behavior in the beginning at a low voltage regime (<0.14 V), above that voltage the *I–V* characteristics shows exponential behavior up to ~2 V, after that the *I–V* retains its linear behavior again due to space charge limited current (SCLC) [31]. The device is rolled many times to examine any possible degradation. The insert shows a photograph of the folded substrate with large-area LEDs. Interestingly, the ZnO NRs-polymer structure was still adhered and no observable changes in the *I–V* characteristics were seen by the rolling/bending process.

The room-temperature EL of the ZnO-polymer NRs LED obtained at an injection current of 1 mA is shown in Fig. 6.6a. This EL reveals a broad emission band covering the visible spectrum from 420 nm to 780 nm. In order to distinguish the contribution from the PFO and ZnO NRs, we used Gaussian fitting to the semi-log EL plot. The blue peak centered at 460 nm belongs to PFO and is attributed to exciton emission and its vibronic progression from non-interacting single chains [32]. This can be explained by the de-convoluted blue peak. It has full width at half maximum (FWHM) around 55 nm and agrees well with the PFO blue peaks, which broaden (428–479 nm) in the emission range of around 50 nm. The broad green emission appearing at 510 nm with a peak centered at 555 nm as shown in Fig. 6.6a is the famous green band emission due to deep level defects in ZnO. This peak has been reported to be in this region, especially in low-temperature-grown ZnO NRs [36], although the source of this green emission is still controversial [4, 37]. The PFO has also a defect emission in this range, i.e., around 500 nm due to the on-chain fluorenone defects and/or electrochemical degradation of the PFO that might be intrinsically introduced during synthesis or generated during the device fabrication and operation [38–40]. The growth of ZnO NRs on top of PFO thin film can also introduce additional defects, which could affect the morphology as well as optical properties of the device as reported for other polymers [41]. The peaks at 640 nm and 710 nm (orange-red and red emission) are frequently attributed to electrons transition within the ZnO wide bandgap, from the zinc interstitial (Zn_i) to deep oxygen interstitial (Oi) site defects [24]. The light emission in the ZnO NRs-polymer hybrid LED suggests that it can be attributed to both the emissions from PFO and to the ZnO deep level defects emissions. The interaction between ZnO NRs surface and PFO molecule makes the PFO link up with ZnO and leads to ZnO NRs-polymer interface. The life time of the electrons at the ZnO NRs-polymer interface is reported to be much larger than that in the ZnO NRs [42].

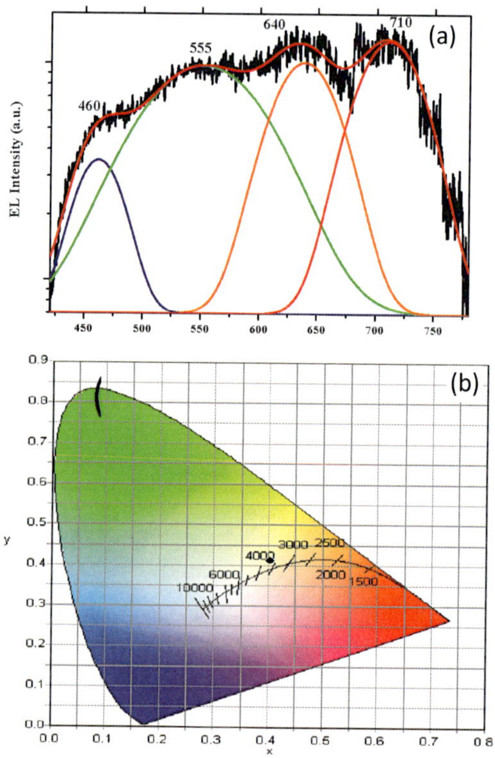

Figure 6.6 (a) RT-EL spectrum obtained at 1 mA injection current and Gaussian fitting of the ZnO nanorod-polymer LED on paper substrate. (b) Chromaticity diagram (CIE coordinates) of the ZnO nanorod-polymer LED on paper substrate.

The electron existing at the interface between the conduction band edge of the ZnO NRs and the lowest unoccupied molecular orbital of the PFO layer slowly drops to the lower states due to their longer life time. As a result, a broad emission ranges from 420 to 780 nm and gives rise to white EL spectrum. The color quality of the present ZnO-PFO LED on paper was investigated with the Commission International DE L'Eclairage (CIE). Figure 6.6b shows that the emitted light has a warm white impression. The color rendering index (CRI) and correlated color temperature (CCT) were found to be 94 and 3660 K, respectively.

6.5 Chemically Grown ZnO Nanorods Printed on Flexible Plastic and Paper Substrates

Zinc oxide (ZnO) has emerged as a very promising material for electronics and optoelectronics applications [36]. It has a rich family of nanostructures (NSs), such as nanorods (NRs) [37], nanotubes [38] and nanowalls [39], which exhibit superior properties in chemical, electric, and optical fields [40]. ZnO can function in a variety of practical applications, including optoelectronic devices, e.g., UV sensors [31], solar cells [41], and LEDs [40, 42].

Several approaches, including very sophisticated and expensive deposition technique such as metal organic chemical vapor deposition (MOCVD) [43], and sputter deposition techniques [44], have been used to synthesize ZnO NRs. These methods are of batch type and difficult to scale up for large-volume production. Alternatives to those highly expensive methods are provided by relatively cheap wet chemical method such as sol-gel [45] and resistive heating, e.g., vapor–liquid–solid (VLS) [46], have been used to synthesize ZnO NSs. However, these processes also yield rather small amount of ZnO material. In most cases, ZnO NRs have been synthesized on substrates but the use of these substrates always limits the scale-up production [47] and normally the synthesis is easier to control in small-scale reactor. A demand for low-cost, large-area devices for lighting requires scalable production of low-cost materials that can be deposited by cheap and scalable methods. As printing and coating are low-cost methods for materials deposition, the objective here is the manufacturing of a printable composition with a high concentration of ZnO NSs. A test print on a semi-automatic screen printer requires at least a few grams to test printability. Here, we have scaled up the preparation of ZnO NRs by a hydrothermal method to produce a few grams per batch. The scale-up is accomplished by increasing the reaction volume and the area onto which ZnO NSs can grow and precipitate. This method is already proven to be successful, and by varying the molar concentration ZnO NRs of various aspect ratios (length/diameter) can be obtained [48].

Flexibility of electronic devices is desirable for both manufacturing and application reasons. A variety of flexible organic electronic components have been already realized. Likewise, flexibility of inorganic electronic devices is desired as well, for a wide range of applications, such as photovoltaic devices [49, 50] flexible batteries [51], and LEDs [52]. While there has been much progress toward the development of flexible organic LEDs, a similar level of maturity has not yet been achieved in flexible inorganic LEDs. One approach to improve the mechanical stability of flexible devices the electronic industry is moving to a different approach such as printed electronics (PE). Printed electronics has a potential to develop new devices, e.g., lightweight flexible LEDs. In this approach, the expensive material and complex method is replaced with a cheap material and low-temperature, low-cost manufacturing techniques, such as inkjet printing, roll-to-roll coasting and screen printing to print devices at large scale. Until now, research has been focused in PE to organic materials, much less effort has been devoted to use inorganic semiconductor. Despite this, some inorganic NSs such as NRs, nanoparticles, and nanowires have an extra advantage for printing; they can be easily fabricated in large amount by solution method, which makes it very suitable for PE [53].

Scaling up the production of surface grown and precipitated ZnO NRs was done by arranging a stack of glass plates supported on a plastic frame immersed in the nutrient growth solution in a 10 L rectangular polypropylene plastic box. The frame for supporting a stack of glass slides was manufactured without using glues, adhesives, or metal connectors, from materials that are cheap and stable to the conditions of ZnO NRs growth, preparation, and washing. Before synthesis, the plastic box, the rack and the glass panes were cleaned with acetone, iso-propanol and deionized water (DI). A seed layer on the glass panes, prepared by the Pacholski et al. method [54] was spray coated on each slide to enhance the density and crystallinity of the ZnO NRs. A 10 L nutrient growth solution containing equimolar (0.075 M) zinc nitrate hexahydrate (Zn $(NO_3)_2 \cdot 6H_2O$) and hexamethylenetetramine (HMT, $C_6H_{12}N_4$) was prepared and after continuous stirring for half an hour it was poured in the plastic box. The rack along with the seeded glass plates was lowered into the nutrient solution and was placed in the ordinary laboratory oven at 95°C for several hours. After terminating the reaction, the growth plates were withdrawn from the vessel and placed in the oven at 100°C until they were dried. Finally, from each growth plate the grown NRs were mechanically scraped by a glass microscopy slide. They were collected, washed with acetone, deionized water several time and then dried. The morphological and structural studies of products were examined at 10 kV using a JEOL JSM-6301F field emission scanning electron microscope (FE-SEM), high-resolution transmission electron microscope (HRTEM), and X-ray diffraction (XRD).

In order to fabricate the LEDs, we have investigated different organic dispersant for ZnO NRs to get a potentially printable ink. After finalizing the organic dispersant, specified amount of ZnO NRs was dispersed in the organic dispersant, yielding a printable ink. A layer of this ink was printed on the patterned substrate layers. The patterned layers were prepared as follows. First, we have printed a pattern of PEDOT: PSS on either paper or plastic substrate and then a poly (9,9-dioctylfluorenyl-2,7-diyl) (PFO) layer was spin coated using spin speed 2000 rpm for 30 sec. For printing the ZnO NRs, we have tested two methods: (1) stencil printing and (2) screen printing. In stencil printing, the ink is placed in a stencil spacer in the shape of adhesive tape defining the area to be printed, and is spread in the space by squeezing with a flat edged object. Stencil print is considered as a first-order printability check, and provides samples for proof of principle.

A screen printing paste was manufactured by dispersing ZnO NRs (5.166 g) in a solvent based screen printing varnish, Plastijet XG 383 (2.5 g, 50%), manufactured by Fujifilm Sericol Ltd, and a small amount (1.1 g) of a retarder, Plastijet ZV 558, was added. The mixture was stirred using an Ultra Turrax™ Tube dispenser before printing.

Screen printing was performed using a nylon web 120-34 (wires per inch and wire diameter in mils) for both electrodes and ZnO NRs paste, using a semiautomatic flat-bed screen printing machine, Screen-printer TIC SCF 550, Eickmeyer. The printed substrates were dried at 90°C in an oven after printing. Finally, we applied aluminum contacts and proceeded for the measurements. Electroluminescence (EL) spectra were acquired on an Andor Shamrock 303iB spectrometer supported with Andor-Newton DU-790 N CCD.

The hydrothermal method is used to perform the synthesis of ZnO NRs. The grown materials from each glass slide were mechanically scraped as shown in Fig. 6.7d. ZnO growth by low-temperature hydrothermal method is substrate independent and produces high-quality NRs. We have used glass slide because in such a large amount solution plastic sheets were not possible to sink in the solution.

After collecting the product, it was carefully cleaned, dried out, and weighted as a result more than 6 g of ZnO NRs was obtained.

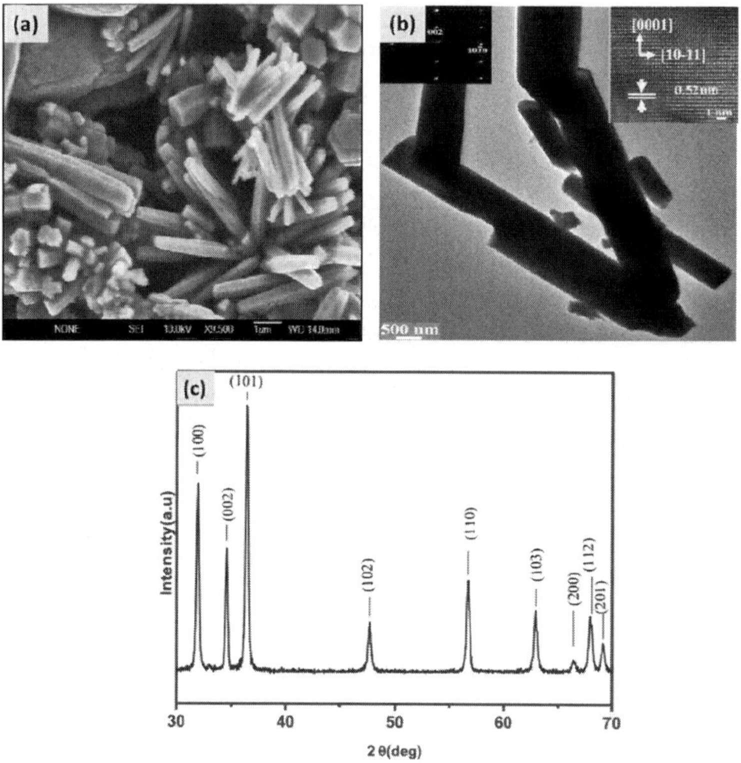

Figure 6.7 (a) shows the SEM image of the ZnO nanorods grown by hydrothemal method at large scale. (b) TEM image of the ZnO nanorods, inset (right) show HRTEM image of the individual nanorods while inset (left) shows the corresponding SAED pattern. (c) X-ray diffraction (XRD) pattern of hydrothermally synthesized ZnO nanorods showing high crystallinity and purity.

The ZnO NRs morphology was first observed using SEM as shown in Fig. 6.7a. All the top surfaces of the grown ZnO NRs were of hexagonal shape with an average diameter of 550 nm, and an average length of 2.2 μm. In this process, the size of the nanorods can be tailored by controlling the hydrothermal conditions such as precursor concentrations, growth temperature, and growth time [48, 50, 55].

An individual ZnO NR was characterized by using the HRTEM and selected area electron diffraction (SAED) to reveal their structural properties (Fig. 6.7b). The image illustrates that the ZnO NR has a high-quality single-crystal structure with a lattice spacing of 0.52 nm and follows [002] growth direction (inset of Fig. 6.7b, right). The corresponding SAED pattern in the inset of Fig. 6.7b (left) insertion is showing that the rod is single crystal and can be readily indexed [002] or *c*-axis growth direction of ZnO NR. This is obvious, since the fastest growth direction for the ZnO nanorods is along the *c*-axis or [001] direction.

The crystallinty of the corresponding product of ZnO NRs was further confirmed by XRD as shown in Fig. 6.7c. The sample for measurement was prepared by spreading the synthesized ZnO nanorods on a substrate, which was then used as a sample holder for XRD. The observed diffraction peaks in the recorded XRD pattern is in good agreement with the values available in the JCPDS card 36-1451 for hexagonal ZnO with wurtzite structure. Usually the relative intensity of peak (002) preferably appear stronger for the case of aligned ZnO NRs grown on a substrate, in our case the XRD pattern slightly differed and resemble to the ZnO powder in which (002) peak weakens while (101) peak intensity appeared stronger. This is because XRD signal originates from the whole sample, which is dominated by the rods randomly distributed and lying down on the substrate. Except ZnO, no other evident diffraction peak was observed, suggesting that the large amount of prepared nanomaterials belongs to high-quality ZnO.

The performance of the ZnO/polymer LEDs is strongly dependent on the processing conditions and environment [56]. Past studies on ZnO/polymer-based devices have relied on ZnO NRs grown within the solution on top of the polymer spin coated substrates [57]. The aqueous nutrient solutions at elevated temperature will degrade the polymer films and as a result the subsequent device performance will be greatly affected [58]. Furthermore, we also used a paper substrate, which cannot survive in the solution environment. Here, we have employed the printed method to avoid aqueous environment in addition to other advantages like large-area printing. The room-temperature electroluminescence (EL) spectrum of the printed LED consisting of PEDOT: PSS/PFO/ZnO NRs/Al is shown in Fig. 6.8a. The EL reveals a broad emission band covering the visible spectrum range from 415 to 630 nm. Two blue shoulder peaks at 450 and

480 nm can clearly be observed in the EL spectrum, the first peak at 450 nm is attributed to the excitonic transition in PFO [59]. The second peak at 480 nm represents a first-order vibronic transition of PFO [59]. On the other hand the EL spectrum demonstrated a broad green emission centered at 532 nm as shown in Fig. 6.8a that can be ascribed to the deep level defects in ZnO. It is known that ZnO NRs grown by the solution method present a large number of defects [60]. The PFO also has defect emission in this range due to the formation of the fluorenone agglomerates or so called keto defect during electro-(photo)oxidation [61]. The reason for this could be when a fluorenone defect is formed, the alkyl chain is lost, favoring local interchain interaction between the PFO backbone [61]. However, the presence of fluorenone defects only may not be sufficient to produce the green band emission. The interchain interactions and cross-linking of the polymers are also proposed to be required for the appearance of the green band [62]. These defects can easily be produced during synthesis and functioning of the device. The emission of the device suggested that the carrier recombination takes place in the PFO/ZnO NRs interface as recently reported [63]. The interaction between the ZnO NRs surface and PFO molecules makes the PFO link up with ZnO. This leads to PFO/ZnO interface producing a strong defect emission [64]. As a result, the device exhibits a broad EL spectrum in the range 415 nm to 630 nm and gives the cold white light as shown in Fig. 6.8a. The color quality of the printed LED was investigated with the Commission International DE L'Eclairage (CIE) as shown in Fig. 6.8b. The resultant light has a cold white impression (bluish) with color coordinate: x = 0.2574 and y = 0.3961 and a correlated color temperature (CCT) of 8367 K.

Using the same low-temperature chemical growth method described above, ZnO nanorods grown on paper substrate were also used as piezoelectric elements to harvest energy from mechanical movements. The word piezoelectricity means "electricity by pressure" and is derived from the Greek "piezin," which means "to press" or "squeeze" and the term "piezoelectric" may have been defined considering the direct piezoelectric effect of certain materials or we can say that piezoelectricity is a form of electricity created when certain crystals are bent or otherwise deformed. The piezoelectric effect, according to the original definition of the phenomenon was first announced during the session of the

Academie des Sciences in Paris on August 2, 1880, by Jacques and Pierre Curie brothers during the studies into the effect of pressure on the generation of electrical charge by crystals (such as quartz) is the ability of the certain crystalline materials to develop an electric charge when those bent or otherwise deformed and the resulting electricity is proportional to a mechanical stress. There are two types of piezoelectric effects: direct piezoelectric effect and indirect or converse piezoelectric effect. The existence of the reverse process was predicted by Lippmann in 1881 and then immediately confirmed by the Curies. In direct piezoelectric effect, when a mechanical force is applied on the crystal, the positive and negative charge centers are initially apparent in dipole moments. So while the overall crystal remains electrically neutral, the difference in charge center displacements results in an electric polarization within the crystal. Electric polarization due to mechanical input is perceived as piezoelectricity.

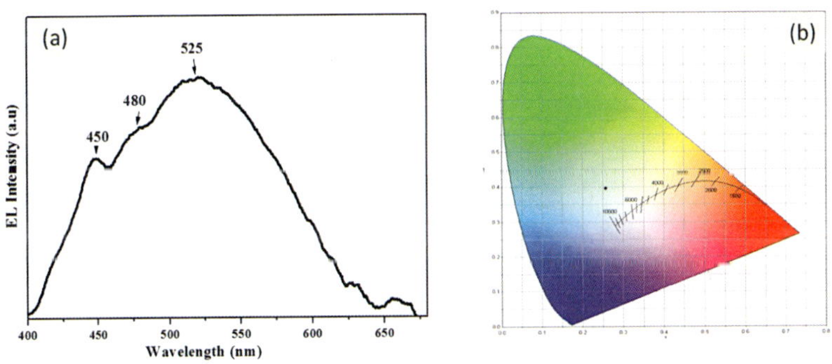

Figure 6.8 (a) The electroluminescence (EL) spectrum of the printed ZnO NR–polymer LEDs. (b) Chromaticity diagram (CIE coordinates) of the printed ZnO NRs-polymer LED on plastic substrate.

ZnO being a wurtzite structure is a most important piezoelectric material with the lack of inversion symmetry it possesses the property of strong piezoelectric effect, which made it as an electromechanical coupling material. The structure of ZnO can be simply described as a number of alternating planes in which each atom is tetrahedrally coordinated and O^{2-} and Zn^{2+} ions stacked alternatively along the *c*-axis and in this situation the center of gravity of the negative charges is at the center of the tetrahedron and due to that positive and negative charges cancel each other but when some force or pressure will be exerted on the crystal along the cornering direction of the tetrahedron, strain produced and the homogeneous macroscopic strain is in general coupled to an internal microscopic strain, i.e., to a relative displacement of the cationic and anionic sublattices. The tetrahedron will experience distortion due to the lattice deformations, and the center of gravity of the negative charges will no longer coincide with the position of the positive central atom. The net effect of force is that an electric dipole is generated. The two opposite faces of the crystal have opposite and unbalance electric charges resulting in a normal dipole moment and atomic-scale spontaneous electric polarization and causes positive and negative charges to migrate to opposite faces of the material (crystal), creating an electric voltage across a piezoelectric voltage that can be used to do work [65]. ZnO has highest effective piezoelectric coefficient among II–IV tetrahedrally coordinated wurtzite compounds semiconductor such as GaN, CdS, AIN, and CdSe. ZnO has the highest piezoelectric tensor and the lack of center of symmetry, combined with a large electromechanical coupling. This results in strong piezoelectric property in ZnO has highest piezoelectric tensor and due to that high piezoelectricity coefficient and electromechanical coupling ZnO attracted to substantial research into the origin of its strong piezoelectricity and applications used in microelectromechanical systems (MEMS) such as piezoelectric transducers [66], piezoelectric pressure sensors [67], piezoelectric actuator and ultrasonic motor [68], resonators, and pass-band filters [69].

For the piezoelectric measurements contact mode of an atomic force microscopy (AFM) having a tetrahedral conductive Si tip coated with platinum (Pt) film was used. There are dual advantages of using a Pt coated tip. First, it increases the conductivity of the electrode and secondly it makes a Schottky contact between the ZnO NWs and the tip, which is critical for the piezoelectric power generation and also plays a major role in the entire process of electron transport, and it directs the current to flow from the electrode to the NWs.

When the tip scans over the surface of the sample, the output signal voltage was continuously monitored. During the scanning of the ZnO NWs, a constant force of about 80 nN is sustained between the tip and the sample and the bending distance of the ZnO NWs was recorded in the topography image as shown in Fig. 6.9. The detailed information regarding the generation of piezopotential can be found elsewhere [70].

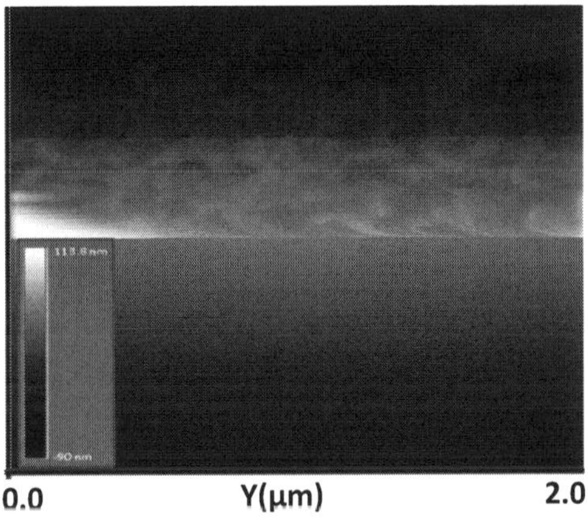

Figure 6.9 Topography image shows the bending distance of ZnO NWs during scanning of the AFM tip on the surface of the paper substrate.

Through this approach, the output voltage can be in the range of a fraction of a volt with some sharp peaks ranging from 4 to 11 mV. The output peaks are corresponding to the maximum deflection of the NWs. A three-dimensional (3D) plot of the output voltage corresponding to the feedback signal from the scanner is shown in Fig. 6.10.

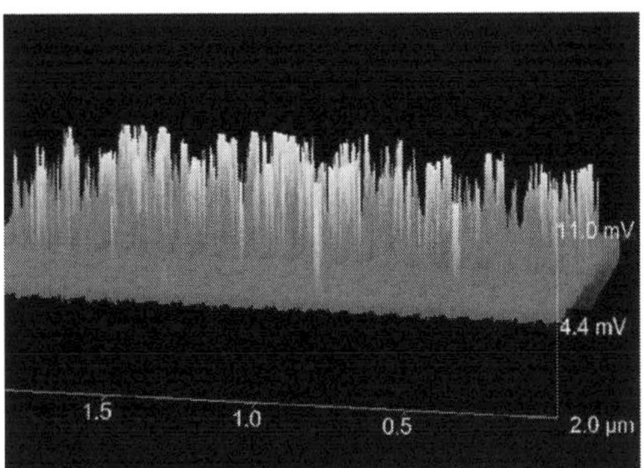

Figure 6.10 Three-dimensional (3D) plot of the output voltage generated during the bending of the ZnO NWs.

Even though the generated potential is not large enough to power a nanodevice, however, this approach provides a convenient way for the future to utilize the piezoelectric properties of ZnO nanowires grown on paper substrate with large enough power for different nanoscale devices. It can be used for generating power used in a broad range of electronic devices without using batteries. This work will provide most significant platform for future soft, portable, and high performance electronic device based on paper substrate. The piezoelectricity produced in ZnO NRs grown on paper substrates by applying the AFM tip but there is a need in future to replace AFM tip with some other alternative external cheap, simple mechanical energy sources such as body movement, Ultrasonic waves, hydraulic energy, which will help the widespread commercialization of the piezoelectricity nanogenerator based on paper substrate.

The combination of ZnO NRs with low-temperature growth process, low-cost, disposable, and flexible paper substrates with cheap mechanical sources for the bending of NRs make it attractive for powering the large-area electronic devices and flexible electronics that can be folded up. Additionally the mechanically flexibility of paper substrate allows it to change shape to meet the space requirement of modern optical devices and dramatically lowers the production costs of electronic devices.

6.6 Conclusion

In summary, we here present a low-temperature solution growth method to synthesize ZnO nanostructures on diverse nonconventional substrates specifically on paper and plastic. This low-temperature solution growth method does not require severe synthesis conditions on the target substrates, which enables the use of paper or plastic substrates that were not possible before for ZnO nanostructures growth. This low-temperature solution growth method has the advantages of simplicity, low cost, nearly 100% growth yield regardless of the synthesis materials and morphologies. The low–temperature solution method will enable electronic devices with desirable properties through their nonconventional substrates, such as flexibility, transparency, adhesion, conductivity, and biocompatibility, and thereby impact a range of applications, such as biosensing, flexible displays, robotics, and energy conversion systems. We believe that the low-temperature solution method can also be used to synthesis other nanomaterials on the nonconventional substrates (e.g., copper oxides, TiO_2, and iron oxides). Printed electronics is an emerging technology with the potential to change the future of technological industry. We can now reliably print basic electronic devices onto paper with cheapest possible routes according to the required output of the device. The UV sensors are fabricated through simple approach of screen printing on paper with pencil-drawn circuitry. Although these sensors are very cheap and easy to fabricate, yet they demonstrate characteristics comparable to those UV sensors that are made with complex and expensive procedures. The printing of the same materials will enable low-cost, large-area ZnO thin-film devices, such as thin-film transistors, thin-film solar cells, and hybrid light-emitting devices on the nonconventional substrates.

References

1. Alvi N, ul Hasan K, Nur O, and Willander M, The origin of the red emission in n-ZnO nanotubes/p-GaN white light emitting diodes, *Nanoscale Res. Lett.*, 2011; **6**, 130.
2. Asif MH, Ali SMU, Nur O, Willander M, Englund UH, and Elinder F, Functionalized ZnO nanorod-based selective magnesium ion sensor for intracellular measurements, *Biosens. Bioelectron.*, 2010; **26**: 1118–1123.
3. Zheng XJ, Cao XC, Sun J, Yuan B, Li QH, Zhu Z, and Zhang Y, A vacuum pressure sensor based on ZnO nanobelt film, *Nanotechnology*, 2011; **22**: 435501.
4. ul Hasan K, Alvi NH, Lu J, Nur O, and Willander M, Single nanowire-based UV photodetectors for fast switching, *Nanoscale Res. Lett.*, 2011; **6**: 348.
5. Martinson ABF, Elam JW, Hupp JT, and Pellin MJ, ZnO nanotube based dye-sensitized solar cells, *Nano Lett.*, 2007; **7**: 2183–2187, 2012/02/15.
6. Zhu G, Yang R, Wang S, and Wang ZL, Flexible high-output nanogenerator based on lateral ZnO nanowire array, *Nano Lett.*, 2010; **10**: 3151–3155, 2012/02/15.
7. Look DC, Recent advances in ZnO materials and devices, *Mater. Sci. Eng. B*, 2001; **80**: 383–387.

8. Wang ZL, ZnO nanowire and nanobelt platform for nanotechnology, *Mater. Sci. Eng. R: Rep.*, 2009; **64**: 33–71.
9. Wang ZL, ZnO nanowire and nanobelt platform for nanotechnology, *Mater. Sci. Eng. R. Rep.*, 2009; **64**: 33–71.
10. Soon-Ku H, Takashi H, Hisao M, Yefan C, Hang-Ju K, Takafumi Y, Akinori T, Hiroyuki S, and Shigeru S, Band alignment at a ZnO/GaN (0001) heterointerface, *Appl. Phys. Lett.*, 2001; **78**: 3349–3351.
11. Novoselov KS, Geim AK, Morozov SV, Jiang D, Zhang Y, Dubonos SV, Grigorieva IV, and Firsov AA, Electric field effect in atomically thin carbon films, *Science*, 2004; **306**: 666–669.
12. Schniepp HC, Li J-L, McAllister MJ, Sai H, Herrera-Alonso M, Adamson DH, Prud'homme RK, Car R, Saville DA, and Aksay IA, Functionalized single graphene sheets derived from splitting graphite oxide, *J. Phys. Chem. B*, 2006; **110**: 8535–8539, 2012/02/16.
13. Castro Neto AH, Guinea F, Peres NMR, Novoselov KS, and Geim AK, The electronic properties of graphene, *Rev. Mod. Phys.*, **81**, 109–162, 2009.
14. Kishwar S, Hasan KU, Tzamalis G, Nur O, Willander M, Kwack HS, and Dang DLS, Electro-optical and cathodoluminescence properties of low temperature grown ZnO nanorods/p-GaN white light emitting diodes, *Phys. Status Solidi A*, 2009; **207**: 67–72.
15. Kishwar S, ul Hasan K, Alvi NH, Klason P, Nur O, and Willander M, A comparative study of the electrodeposition and the aqueous chemical growth techniques for the utilization of ZnO nanorods on p-GaN for white light emitting diodes, *Superlattices Microstruct.*, 2010; **49**: 32–42.
16. Fulati A, Ali SMU, Asif MH, Alvi NUH, Willander M, Brännmark C, Strålfors P, Börjesson SI, Elinder F, and Danielsson B, An intracellular glucose biosensor based on nanoflake ZnO, *Sens. Actuators B: Chem.*, 2010; **150**: 673–680.
17. Wang M, Ye C-H, Zhang Y, Hua G-M, Wang H-X, Kong M-G, and Zhang L-D, Synthesis of well-aligned ZnO nanorod arrays with high optical property via a low-temperature solution method, *J. Cryst. Growth*, 2006; **291**: 334–339.
18. Zainelabdin A, Zaman S, Amin G, Nur O, and Willander M, Deposition of well-aligned ZnO nanorods at 50°C on metal, semiconducting polymer, and copper oxides substrates and their structural and optical properties, *Cryst. Growth Des.*, **10**, 3250–3256, 2011/12/26.
19. Scrosati B, Nanomaterials—Paper powers battery breakthrough. *Nat. Nanotechnol.*, 2007; **2**: 598–599.

20. Graham-Rowe D, Electronic paper rewrites the rulebook for displays. *Nat. Photon.*, 2007; **1**: 248–251.

21. Ozgur U, Alivov YI, Liu C, Teke A, Reshchikov MA, Dogan S, et al., A comprehensive review of ZnO materials and devices. *J. Appl. Phys.*, 2005; **98**, 041301.

22. Willander M, Nur O, Zhao QX, Yang LL, Lorenz M, Cao BQ, et al., Zinc oxide nanorod based photonic devices: recent progress in growth, light emitting diodes and lasers. *Nanotechnology*, 2009; **20**, 332001.

23. Travlos A, Boukos N, Chandrinou C, Zinc related defects in ZnO nanorods. *Phys. Status Solidi B*, 2012; **249**: 560–563.

24. Amin G, Hussain I, Zaman S, Bano N, Nur O, Willander M, Current-transport studies and trap extraction of hydrothermally grown ZnO nanotubes using gold Schottky diode. *Phys. Status Solidi* A, 2010; **207**: 748–752.

25. Zainelabdin A, Zaman S, Amin G, Nur O, Willander M, Deposition of well-aligned ZnO nanorods at 50°C on metal, semiconducting polymer, and copper oxides substrates and their structural and optical properties. *Cryst. Growth Des.*, 2010; **10**: 3250–3256.

26. Vayssieres L, Growth of arrayed nanorods and nanowires of ZnO from aqueous solutions. *Adv. Mater.*, 2003; **15**: 464–466.

27. Zainelabdin A, Zaman S, Amin G, Nur O, Willander M, Deposition of well-aligned ZnO nanorods at 50°C on metal, semiconducting polymer, and copper oxides substrates and their structural and optical properties. *Cryst. Growth Des.*, 2010; **10**: 3250–3256.

28. Alvi NH, Riaz M, Tzamalis G, Nur O, Willander M, Fabrication and characterization of high-brightness light emitting diodes based on n-ZnO nanorods grown by a low-temperature chemical method on p-4H-SiC and p-GaN. *Semicond. Sci. Tech.*, 2010; **25**, 065004.

29. Alvi NH, Usman Ali SM, Hussain S, Nur O, Willander M, Fabrication and comparative optical characterization of n-ZnO nanostructures (nanowalls, nanorods, nanoflowers and nanotubes)/p-GaN white-light-emitting diodes. *Scripta Mater.*, 2011; **64**: 697–700.

30. Chen LJ, Manekkathodi A, Lu MY, Wang CW, Direct growth of aligned zinc oxide nanorods on paper substrates for low-cost flexible electronics. *Adv. Mater.*, 2010; **22**: 4059–4063.

31. Seminario JM, Gimenez AJ, Yanez-Limon JM, ZnO-paper based photoconductive UV sensor. *J. Phys. Chem. C*, 2011; **115**: 282–287.

32. Lee CH, Kim DR, Zheng XL, Fabrication of nanowire electronics on nonconventional substrates by water-assisted transfer printing method. *Nano Lett.*, 2011; **11**: 3435–3439.

33. Kind H, Yan H, Messer B, Law M, and Yang P, Nanowire ultraviolet photodetectors and optical switches, *Adv. Mater.*, 2002; **14**: 158–160.
34. Soci C, Zhang A, Xiang B, Dayeh SA, Aplin DPR, Park J, Bao XY, Lo YH, and Wang D, ZnO nanowire UV photodetectors with high internal gain, *Nano Lett.*, 2007; **7**: 1003–1009.
35. Bai S, Wu W, Qin Y, Cui N, Bayerl DJ, and Wang X, High-performance integrated ZnO nanowire UV sensors on rigid and flexible substrates, *Adv. Funct. Mater.*, 2011; **21**: 4464–4469.
36. Ozgur U, Alivov YI, Liu C, Teke A, Reshchikov MA, Dogan S, et al., A comprehensive review of ZnO materials and devices. *J. Appl. Phys.*, 2005; **98**, 041301.
37. Ko SH, Lee D, Kang HW, Nam KH, Yeo JY, Hong SJ, et al., Nanoforest of hydrothermally grown hierarchical ZnO nanowires for a high efficiency dye-sensitized solar cell. *Nano Lett.*, 2011; **11**: 666–671.
38. Zhou JZ, Guo HH, Lin ZH, Feng ZF, Lin LL. White-light-emitting diode based on ZnO nanotubes. *J. Phys. Chem. C*, 2009; **113**: 12546–12550.
39. Israr MQ, Sadaf JR, Nur O, Willander M, Salman S, Danielsson B. Chemically fashioned ZnO nanowalls and their potential application for potentiometric cholesterol biosensor. *Appl. Phys. Lett.*, 2011; **98**: 253705–253703.
40. Willander M, Nur O, Zhao QX, Yang LL, Lorenz M, Cao BQ, et al., Zinc oxide nanorod based photonic devices: recent progress in growth, light emitting diodes and lasers. *Nanotechnology*, 2009; **20**: 332001.
41. Wei M, Han JB, Fan FR, Xu C, Lin SS, Duan X, et al., ZnO nanotube-based dye-sensitized solar cell and its application in self-powered devices. *Nanotechnology*, 2010; **21**, 405203.
42. Zhang Y, Zhang XM, Lu MY, Chen LJ, Wang ZL, Fabrication of a high-brightness blue-light-emitting diode using a ZnO-nanowire array grown on p-GaN thin film. *Adv. Mater.*, 2009; **21**: 2767–2770.
43. Wu JJ, Wong DKP, Fabrication and impedance analysis of n-ZnO nanorod/p-Si heterojunctions to investigate carrier concentrations in Zn/O source-ratio-tuned ZnO nanorod arrays. *Adv. Mater.*, 2007; **19**: 2015–2019.
44. Ting JM, Chen MT, Sputter deposition of ZnO nanorods/thin-film structures on Si. *Thin Solid Films*, 2006; **494**: 250–254.
45. Shi SB, Xu JP, Zhang XS, Li L, Effect of annealing on the structural and luminescent properties of ZnO nanorod arrays grown at low temperature. *J. Appl. Phys.*, 2011; **109**, 103508.

46. Zhao QX, Klason P, Willander M, Growth of ZnO nanostructures by vapor-liquid-solid method. *Appl. Phys. Mater. Sci. Proc.*, 2007; **88**: 27–30.

47. Yuan FL, Peng H, Bai LY, Li JL, Chen YF, Plasma synthesis of large quantities of zinc oxide nanorods. *J. Phys. Chem. C*, 2007; **111**: 194–200.

48. Wang ZL, Xu S, Lao C, Weintraub B, Density-controlled growth of aligned ZnO nanowire arrays by seedless chemical approach on smooth surfaces. *J. Mater. Res.*, 2008; **23**: 2072–2077.

49. Rogers JA, Baca AJ, Yu KJ, Xiao JL, Wang SD, Yoon J, et al., Compact monocrystalline silicon solar modules with high voltage outputs and mechanically flexible designs. *Energy Environ. Sci.*, 2010; **3**: 208–211.

50. Yu H, Zhang Z, Han M, Hao X, Zhu FA, General low-temperature route for large-scale fabrication of highly oriented ZnO nanorod/nanotube arrays. *J. Am. Chem. Soc.*, 2005; **127**: 2378–2379.

51. Gaikwad AM, Whiting GL, Steingart DA, Arias AC, Highly flexible, printed alkaline batteries based on mesh-embedded electrodes. *Adv. Mater.* 2011; 23: 3251–3255.

52. Zainelabdin A, Zaman S, Amin G, Nur O, Willander M, Stable white light electroluminescence from highly flexible polymer/ZnO nanorods hybrid heterojunction grown at 50°C. *Nanoscale Res. Lett.*, 2010; **5**: 1442–1448.

53. Toyama T, Takeuchi H, Yamaguchi D, Kawasaki H, Itatani K, Okamoto H, Solution-processed ZnO nanocrystals in thin-film light-emitting diodes for printed electronics. *J. Appl. Phys.*, 2010; **108**, 084302.

54. Pacholski C, Kornowski A, Weller H, Self-assembly of ZnO: from nanodots to nanorods. *Angew. Chem. Int. Ed. Engl.*, 2002; **41**: 1188–1191.

55. Wang M, Ye CH, Zhang Y, Hua GM, Wang HX, Kong MG, et al., Synthesis of well-aligned ZnO nanorod arrays with high optical property via a low-temperature solution method. *J. Cryst. Growth*, 2006; **291**: 334–339.

56. Olson DC, Shaheen SE, Collins RT, Ginley DS, The effect of atmosphere and ZnO morphology on the performance of hybrid poly (3-hexylthiophene)/ZnO nanofiber photovoltaic devices. *J. Phys. Chem. C*, 2007; **111**: 16670–16678.

57. Bolink HJ, Coronado E, Sessolo M, White hybrid organic-inorganic light-emitting diode using ZnO as the air-stable cathode. *Chem. Mater.*, 2009; **21**: 439–441.

58. Olson DC, Lee YJ, White MS, Kopidakis N, Shaheen SE, Ginley DS, et al., Effect of polymer processing on the performance of poly (3-hexylthiophene)/ZnO nanorod photovoltaic devices. *J. Phys. Chem. C*, 2007; **111**: 16640–16645.

59. Wadeasa A, Nur O, Willander M, The effect of the interlayer design on the electroluminescence and electrical properties of n-ZnO nanorod/p-type blended polymer hybrid light emitting diodes. *Nanotechnology*, 2009; **20**: 065710.

60. Djurisic AB, Leung YH, Optical properties of ZnO nanostructures. *Small*, 2006; **2**: 944–961.

61. Montilla F, Mallavia R, On the origin of green emission bands in fluorene-based conjugated polymers. *Adv. Funct. Mater.*, 2007; **17**: 71–78.

62. Zhao W, Cao T, White JM, On the origin of green emission in polyfluorene polymers: The roles of thermal oxidation degradation and crosslinking. *Adv. Funct. Mater.*, 2004; **14**: 783–790.

63. Wadeasa A, Tzamalis G, Sehati P, Nur O, Fahlman M, Willander M, et al., Solution processed ZnO nanowires/polyfluorene heterojunctions for large area lightening. *Chem. Phys. Lett.*, 2010; **490**: 200–204.

64. Lee CY, Wang JY, Chou Y, Cheng CL, Chao CH, Shiu SC, et al., White-light electroluminescence from ZnO nanorods/polyfluorene by solution-based growth. *Nanotechnology*, 2009; **20**, 425202.

65. Service, RF, Nanogenerators tap waste energy to power ultrasmall electronics, *Science*, 2010; **328**: 304–305.

66. Minne, SC, Manalis, SR, Quate CF, Parallel atomic force microscopy using cantilevers with integrated piezoresistive sensors and integrated piezoelectric actuators, *Appl. Phys. Lett.*, 1995; **67**: 3918.

67. Choi, JR, Polla, Dennis. L, Integration of microsensors in GaAsI MESFET process, *J. Micromechan. Micro-Eng.*, 1993; **3**: 60–64.

68. Uchino K, Piezoelectric ultrasonic motors: overview, *Smart Mater. Struct.*, 1998; **7**: 273–285.

69. Molarius, J, Kaitila, J, Pensala, T, Ylilammi, M, Piezoelectric ZnO films by r.f sputtering, *J. Mater. Sci. Mater. Electron.*, 2003; **14**: 431–435.

70. Wang, ZL, Song, J, Piezoelectric nanogenerators based on zinc oxide nanowire arrays, *Science*, 2006; **312**: 242–246.

Index

A-excitons 45
absorbance 166
absorption 33, 147, 150
acetone 198
AFM, *see* atomic force microscopy
ALD, *see* atomic layer deposition
aligned zinc oxide nanowire arrays 102–3, 111, 121
alloys, ternary 45, 53, 55
alternating current 118
anions 4–5, 105
applications
  biological 160–61
  biomedical 158, 170, 190
  optical 82–83, 128
  opto-electronics 191
  sensing 129
atomic force microscopy (AFM) 101, 107, 176, 189, 204
atomic layer deposition (ALD) 3, 11

bandgap energies 49–50, 52–53, 55, 80
BEM, *see* boundary element method
biocompatibility 158, 160, 178, 206
biological analytes 127, 129–30, 142–43, 145, 151
biosensors 129, 142, 144, 146, 158, 186–87
boundary element method (BEM) 63
Bragg mirrors 67–68
cadmium 53–55, 58–60
cadmium atoms 52, 55
calcium ions 133, 136–37, 150
cancer cell necrosis 148, 158
cancer cells 147–48, 170
cancer therapy 170
capacitors 120, 122–23
catalysator 5, 7
catalyst 14–15, 17, 22
catalytic growth 15
cavity effects 43–44, 60–61, 63, 65, 67, 69
CBD, *see* chemical bath deposition
CBD-grown zinc oxide nanopillars 32
CBD growth 6–7, 16
CBD growth rates on FTO substrates 8, 16
CCT, *see* correlated color temperature
cell culture 162–63
cell death 133, 147, 159–60, 168, 170, 173–74, 178
cell killing mechanism 171, 180–81
cell lines 159, 174
  malignant 159, 161, 174–75, 178
cell necrosis 149–50, 178, 180
cell viability 165, 167, 170, 173–74, 181
cells
  adipocytes 134, 137, 140
  cancerous 147, 181
  malignant 164–65
  melanoma 178, 182
  normal 157, 182

chemical bath deposition (CBD) 3–6, 16, 25, 34
chemical growth
  low-temperature 128, 130, 142
  wet 151
chemical vapor deposition (CVD) 3, 16, 34, 59
chemical vapor transport (CVT) 3, 13, 34
cholesterol 142–44
cholesterol detection 142
cholesterol oxidase 142–43
cholesterol sensing using zinc oxide nanostructures 142
color rendering index (CRI) 196
conductivity 78, 204, 206
confocal microscopy 167, 169, 171, 173
copper wires 131
core–shell heterostructures 57
core–shell nanostructures 4, 32–33
  spherical Zn/ZnO 73
correlated color temperature (CCT) 196, 202
CRI, *see* color rendering index
CVD, *see* chemical vapor deposition
CVT, *see* chemical vapor transport
cyclotene 188
cytotoxicity 161, 164, 169–70, 174–75, 178

D-gluconolactone 139
density functional theory (DFT) 49–50
DFT, *see* density functional theory
direct current 101, 110–11, 113, 115
DMEM, *see* Dulbecco's modified eagle medium
drug delivery
  freestanding 158, 164
  microinjection 158, 175
  targeted 158, 161
Dulbecco's modified eagle medium (DMEM) 162, 165, 167

elastic modulus 110
electric polarization 203
electricity 102, 104, 106, 111, 123–24, 202
electrochemical sensors 130
electrode 8, 111–12, 118–19, 133–34, 136, 144–45, 151, 186, 189, 199, 204
  nanostructured 24
  zigzag 112–13, 115
electrodeposition 3, 8, 10–11
electroluminescence 75–76, 199, 203
electronic devices 190, 197, 206
  disposable 190–91
electronic paper displays (EPDs) 190
electronics, printed 198, 207
electrons 25, 78, 118, 195–96
  unpaired 78, 193
emission 27, 31, 45–47, 49, 57–58, 67, 75, 149, 195, 202
emission energies 52–56, 58, 60
emission origins 57–58
energy
  electrical 106, 108, 128–29
  photon 77–78
EPDs, *see* electronic paper displays
excitation, fundamental 81
excitation intensity 66–67, 79
exciton emission, free 45–47, 50–51
excitonic emissions 49
exciton–polaritons 67
excitons 49–51, 60, 72
  free 46–47, 50–51

Fabry–Perot modes 43–44, 60, 65, 82
FBS, *see* fetal bovine serum
FDTD, *see* finite difference time domain
FDTD simulations 63–64
fetal bovine serum (FBS) 162
fibroblast cells, foreskin 178, 181
finite difference time domain (FDTD) 63
fluorenone defects 202
fluorescence 149, 167–69, 178
foreskin fibroblast 158, 162, 164, 174–75, 178–79
free standing zinc oxide nanowire core cavities 68

GaN 2, 14, 26, 28–31, 191, 204
  growth of 29, 31
  p-doped 26, 28
GaN-based nanostructures 31
GaN layers 29
GaN nanostructures 29
GaN overgrowth of nanostructures 31
gas-phase growth 3–4
generation, harmonic 43
generator, lateral wire 118, 120
glass substrates 12, 23–24, 32–33
gluconic acid 139
glucose 129, 138–39, 141, 150
glucose detection 150–51
glucose measurements 139
glucose oxidase 139
glucose uptake, insulin-induced activation of 141
gold 8, 15
gold electrodes 121–22
gold particles, small 103
graphene-pencil analogy 187
graphite 187–88

hepatocellular carcinoma 158, 162, 167, 181–82
heterostructures 3, 52, 55–56, 59–60, 80
  nanowire 45, 56
high-resolution transmission electron microscope (HRTEM) 199, 201
HRTEM, *see* high-resolution transmission electron microscope
human body 129, 132, 138, 142, 144
hydrothermal method 191, 194, 197, 199

incubation 161, 165, 167
induction 174
infrared spectroscopy 43–44, 82
insulin 141
irradiation 147, 163, 165–66, 169, 178–79
isoelectric points 142

Jablonski diagram 159–60

laser irradiation 161, 165
lasers 44, 75, 128, 149
lasing 66, 77
  random 43–44, 60, 68–69, 82
lasing threshold 65–66
lateral wire generator (LWG) 118–20
LDL, *see* low-density lipoprotein
LEDs, *see* light-emitting diodes
light-emitting diodes (LEDs) 17, 26, 44, 51, 186–87, 197–99

light waves, constructive interference of 61–62
linear motor 123
liquid phase epitaxy (LPE) 3
low-density lipoprotein (LDL) 169
LPE, *see* liquid phase epitaxy
luminescence measurements 44–45, 51
LWG, *see* lateral wire generator

magnetic resonance imaging (MRI) 158
MBE, *see* molecular beam epitaxy
mechanical deformation 111, 123
melanocytes 161, 178, 181–82
melanoma 158, 161–62, 174–75
metal organic chemical vapor deposition (MOCVD) 16–17, 26, 29, 31–32, 34–35, 197
metal organic vapor phase epitaxy (MOVPE) 53, 56, 58
metal oxide nanostructures 164
  preparation of 164
metallic ions 136, 138
microscope
  epifluorescence 164, 166–67
  optical 76, 107–8
MOCVD, *see* metal organic chemical vapor deposition
molecular beam epitaxy 3
molecular beam epitaxy (MBE) 3, 17, 25, 34–35, 53, 55
MOVPE, *see* metal organic vapor phase epitaxy
MQWs, *see* multiple quantum wells
MRI, *see* magnetic resonance imaging
multiple quantum wells (MQWs) 25, 59

n-ZnO nanostructures 27
nanogenerator 101–7, 110–15, 120, 123, 125
  direct current 101, 110–11, 113, 115, 120, 124
  laterally packed AC 118–19
  high-power-output 115, 117
  high-output lateral 120–21, 123
nanoheterostructures 12, 43–44, 82
nanomaterials 158–59, 164–66, 170, 174, 178, 207
nanoparticles 24, 158, 170, 174, 186, 190, 198
nanophotonic devices 81–82
nanopillars 3, 6–9, 14–15, 18–20, 28, 32, 35
  growth of 15, 19
  high-quality 1–2
nanopowders 24, 35
nanorods 10, 20, 22, 25, 44, 127, 158, 186, 191, 197, 200
nanostructure-based devices 22
nanostructure devices 75, 77, 79, 81
nanostructure fabrication 15, 34
nanostructure growth 21
nanostructure morphology 49
nanostructure synthesis 190
nanostructures 2–4, 7, 11, 26, 29, 31, 34–35, 43–45, 50–52, 70–73, 127–29, 175–79, 181, 185–86, 190
  advanced QW 26
  implementation of 1–2
  one-dimensional 158
nanotubes 106, 130, 132, 158, 186, 197
nanowalls 19–20, 144, 197
nanowire ensembles 74
nanowires 10, 15–16, 20, 32, 44, 52, 66, 74–75, 77, 80, 127, 144, 151, 158, 186

long 7
tapered 62, 66
NBE, *see* near-band edge
near-band edge (NBE) 45
near-surface light wave manipulation 44
neutral red assay (NRA) 165
nonconventional substrates 185, 191, 206–7
NRA, *see* neutral red assay

oocytes 133, 139–40
oocytes frog cells 137
optical recombination 43–45, 47, 49, 51, 53, 55, 57, 59, 83
optoelectronic applications 2
optoelectronic devices 56, 82
optoelectronics applications 197
organic additives 22
organic dispersant 199
oxygen 17, 20, 77–78, 181, 193
molecular 147
singlet 147, 150, 178
oxygen molecules 78, 192

p-type ZnO 76
p-type ZnO nanowire arrays 125
paper substrates 185–86, 190–91, 193–94, 196–97, 199, 201–3, 205–6
bare 188
coated 188
flexible 206
pre-coated 194
PBS, *see* phosphorous buffered solution
PFW, *see* piezoelectric fine wire
phonon modes 74
phonons 70, 72–73
phosphorous buffered solution (PBS) 144–45, 164–66, 176
photochemical reactions 159, 179
photodetectors 43–44, 77, 82, 192
photodynamic therapy 82, 147–48, 157, 169
photoluminescence 69
photosensitizers 147, 158, 164–67, 169–70, 172–73, 175, 181–82
photovoltaic applications 4, 8, 16, 22, 24, 33
photovoltaic devices 1, 198
physical vapor transport (PVT) 3, 13, 34
piezoelectric coefficient 110
piezoelectric effect 105, 202–3
direct 202–3
piezoelectric fine wire (PFW) 118–19
piezoelectric materials 82, 158
piezoelectric nanogenerator based on zinc oxide nanomaterials 101–2, 104, 106, 108, 110, 112, 114, 116, 118, 120, 122, 124
piezoelectric tensor, highest 204
piezoelectric voltage 118, 204
piezoelectricity 128, 186, 202–3, 206
plastic substrates 35, 199, 203, 206
platinum 15, 106, 111, 204
PLD, *see* pulsed laser deposition
polariton lasing 67
polarization 71, 81
porosity 22, 24
power generation 101, 103, 107, 111–12, 124, 186
printed electronics 198, 207
printing
screen 188, 198–99, 207

stencil 199
pulsed flow 29–30
pulsed laser deposition (PLD) 3, 21–22, 28, 34–35, 55, 57–59, 190
pulsed precursor flows 29
pulses, electrical 104, 106, 108
PVT, *see* physical vapor transport

QDs, *see* quantum dots
quantum dots (QDs) 43, 45, 52, 55, 59
quantum wells (QWs) 2, 45, 55–56, 58–59
quantum wires 49–50
QWs, *see* quantum wells
radial 57–58

random lasers 68
reactive ion etching (RIE) 194
reactive oxygen species (ROS) 159–60, 167, 178–80
reflection high-energy electron diffraction (RHEED) 25
resonators 62, 129, 158, 204
RHEED, *see* reflection high-energy electron diffraction
RIE, *see* reactive ion etching
ROS, *see* reactive oxygen species

SAED, *see* selected area electron diffraction
sapphire 19, 111
plane 17, 20–21
sapphire substrate 28, 102–3
scanning electron microscope (SEM) 30, 131–32, 188–89, 199
Schottky barrier 79, 102, 106, 108, 112
SCLC, *see* space charge limited current
second-harmonic generation (SHG) 80–82
selected area electron diffraction (SAED) 201
SEM, *see* scanning electron microscope
semiconductors 13, 26, 72–73, 129, 158, 186
sensing devices 151
sensor fabrication 4, 12
sensors 4, 9, 24, 127, 129–31, 150–51, 186, 193, 207
nanostructure-based 127, 129
nanostructured zinc oxide 12
potentiometric 130
SHG, *see* second-harmonic generation
SILAR, *see* successive ion layer adsorption reaction
SILAR-grown core–shell nanostructures on zinc oxide nanorods 32
SILAR-grown Cu_2O nanowires 32–33
silicon substrate 8–9, 15, 74, 107
silver paste 103, 106–7
silver wire 143, 150
sodium ions 136
solar cells 4, 11, 32, 186, 197
dye-sensitized 8, 16, 24, 82
space charge limited current (SCLC) 195
spin-coating 22
SPR, *see* surface plasmon resonance
substrate fabrication technology 3
substrate temperature 13–15, 17–20, 31, 34
successive ion layer adsorption reaction (SILAR) 3, 12, 32, 34

surface excitons 46, 48
surface plasmon resonance (SPR) 160

TEM, *see* transmission electron microscopy
thin films 2, 11–12, 44–45, 47, 52, 54–55, 60, 69, 74–75, 80, 106, 117, 128, 190, 195
transducers, physiochemical 130
transistors 9, 150
transmission electron microscopy (TEM) 189
tumors 147, 170

ultra-violet (UV) 33, 79–80, 149, 161, 181, 186, 191
ultrasonic wave 106, 111–15, 117, 206
uric acid 142, 144–46
uric acid biosensor 144, 146
uric acid detection using zinc oxide nanowires 144
urine 144
UV, *see* ultra-violet
UV exposure 149, 178, 193
UV illumination 77, 192–93
UV lamp 192
UV light 70, 158, 167, 169–70, 174–75, 178–80, 192
UV sensors 186–87, 192, 197, 207

vapor deposition method 121
vapor phase transport (VPT) 13, 31, 34–35, 53
VPT, *see* vapor phase transport
wet chemical deposition 4, 33
wet chemical fabrication 4–5, 7, 9, 11
wet chemical method 115–16
WGM, *see* whispering gallery modes
WGM lasing in zinc oxide nanostructures 64
whispering gallery modes (WGM) 60–66, 81
wireless devices 129
wurtzite structures 70, 73, 201, 203

X-ray diffraction (XRD) 189, 199–201
XRD, *see* X-ray diffraction

zinc oxide 1–2, 12, 128, 158–59, 170, 185–86, 197
  bulk 58, 72–73, 77
  nanostructured 4, 12, 72
zinc oxide–based core–shell nanostructures 12, 32
zinc oxide–based ternary alloys 52–53
zinc oxide bulk-substrate 29–30
zinc oxide device applications 3
zinc oxide epitaxial growth 3
zinc oxide epitaxial layer quality 1–2
zinc oxide growth 5, 11, 15, 22, 27, 128, 199
zinc oxide heteronanostructures 2
zinc oxide nano-crystals 186
zinc oxide nano-tetrapods 81
zinc oxide nanobelts 70–71
zinc oxide nanocrystals 5–6
  screen-printed 187
zinc oxide nanoflakes 177
zinc oxide nanomaterials 159, 170
zinc oxide nanoparticles 69, 80, 157, 159–60, 169, 174

capped 169
zinc oxide nanopillars 7–8, 14, 16, 25–26, 28–29, 31–35
high-density 5
long 16
self-organized 22
zinc oxide nanoplate-nanowire architectures 44
zinc oxide nanopowder approach 24
zinc oxide nanopowder-based technologies 22–23
zinc oxide nanopowder layers 23
zinc oxide nanopowders 2, 4, 22–23
zinc oxide nanorods 157, 159
zinc oxide nanowires 147, 149, 159
zinc oxide-based heteronanostructures 25, 27, 29, 31
zinc oxide nanorod arrays 81
zinc oxide nanorod buffer 25
zinc oxide nanorod fabrication methods 34
zinc oxide nanorod intracellular foreskin fibroblast 180
zinc oxide nanorod tip 181
zinc oxide nanorods 4, 8, 10, 33, 143, 157–60, 164, 166–68, 170, 172–76, 178–82, 187–88, 194, 196–202, 206
aligned 201
as-grown 143
bare 167
dispersing 199
foreskin fibroblast intracellular 180
grown 143
growth of 175, 188, 193, 195, 198
high-density 5–6
individual 201
intracellular 180
low-temperature-grown 195
pumped 44
single 71
synthesized 200–1
zinc oxide nanorods on paper substrates 190
zinc oxide nanostructure fabrication methods 3
zinc oxide nanostructure fabrication techniques 2
zinc oxide nanostructure fabrication technologies 2
zinc oxide nanostructure growth 22, 185, 206
zinc oxide nanostructures
fabrication of 1–6, 8, 10, 12, 14, 16–18, 20, 22, 24, 26, 28, 30, 32–34
optical recombination of 45
pulsed laser deposition of 22
zinc oxide nanotechnology development 35
zinc oxide nanowire arrays 34
zinc oxide nanowire fabrication 15
zinc oxide nanowire sensors 150
zinc oxide nanowire structures 75, 77
zinc oxide nanowire tetrapod 81
zinc oxide nanowires 20, 43–44, 67, 78, 81–82, 102–4, 110, 124–25, 130–33, 138–39, 142, 146–47, 149–51, 175–76, 204–6
aligned 74
as-synthesized 121
bare 149
conjugated 149
excited 66, 80
fabrication 13
first device applications of 44
functionalized 151
hexagonal 131

n-type 76, 106
semiconducting 104–5
thin 80
zinc oxide powder 21, 69, 201
zinc oxide substrates 29, 31
zinc oxide wire 107–8, 118